T0226151

Practical Insecurity: The Layman's Guide to Digital Security and Digital Self-defense

Published 2023 by River Publishers

River Publishers

Alsbjergvej 10, 9260 Gistrup, Denmark

www.riverpublishers.com

Distributed exclusively by Routledge

605 Third Avenue, New York, NY 10017, USA

4 Park Square, Milton Park, Abingdon, Oxon OX14 4RN

Practical Insecurity: The Layman's Guide to Digital Security and Digital Self-defense / by Lyndon Marshall.

Routledge is an imprint of the Taylor & Francis Group, an informa business

ISBN 978-87-7022-989-0 (paperback)

ISBN 978-10-0098-438-5 (online)

ISBN 978-1-003-45302-4 (ebook master)

A Publication in the River Publishers series
RAPIDS SERIES IN COMPUTING AND INFORMATION SCIENCE AND TECHNOLOGY

Practical Insecurity: The Layman's Guide to Digital Security and Digital Self-defense

Lyndon Marshall

University of Providence, USA

Routledge
Taylor & Francis Group

NEW YORK AND LONDON

Contents

Preface

Cybersecurity is of increasing importance to all of us. We are constantly bombarded with news regarding the latest hack into some organization, the development of some new computer virus, the exposure of someone's data somewhere, the latest attack by China or Russia or Iran, the latest flaw in software used by millions, the most recent new form of digital blackmail, and so on. We live in fear of our lives being stolen digitally. Yet, outside of the constant injunctions to keep ourselves and our data safe, we are given very little in terms of how we can protect ourselves.

The unfortunate part of how things go is the fact that the great preponderance of books on technology fall into two areas: teaching technologists how to better do their job or teaching the complete novice how to use a particular piece of software. There is an unfortunate gap in the literature available on computing. Nobody seems interested in teaching the average person how to protect themselves in the increasingly murky digital forest. This form of digital self-protection is what this book is about.

This book has its origins in the time in which I was CTO for a small university. In addition to very substantial network engineering work, I got to see the results of the digital antics of end-users on our network, and I got to hear about people's digital problems. In some very significant ways, this book is a distillation of what I've seen, what I've heard, and the advice I've given to users over the years.

Regard this book as a digital self-defense manual. Reading it and practicing what it recommends is no guarantee that you won't be hacked or suffer the acquisition of a virus, just like getting a black-belt in a martial art is no guarantee that you won't get mugged. However, if you put the ideas in this book into practice and use it to help guide your thinking about your own digital security, this book will help you to reduce the level of risk you face when using computers, tablets, laptops, or whatever form of digital tech that you like to indulge in.

Read this book while asking yourselves what you can do to instill in your use of technology the ideas and methods proposed in this manual. I hope you enjoy this book and that you find the ideas offered herein to be useful.

Lyndon Marshall

About the Author

Lyndon Marshall has worked in a wide variety of technical areas. He has been a software engineer, a systems analyst, a project manager, a network administrator, a network engineer, a market research analyst, and a professor of Computer Science. For many years, he taught Computer Science at the University of Providence where he taught subjects such as computer programming, systems analysis, database systems, computer architecture and organization, and computer security. His courses in Computer Security included Principles of Computer Security, Computer Forensics, and Penetration Testing. Currently, he is teaching programming, database systems, and neural networks as part of the Applied Mathematics program at the University of Providence.

1

The Security Landscape

First, let's talk about what this book is and is not. This book is not a "how to use" book. Software and tools evolve on a daily basis. Folks who devote themselves to communicating how a particular program works perform a wonderful service, but that is not what this book is about. It is about thinking in ways that will help protect you and your data.

Additionally, this book is not for security pros. Most will find the information in here to be pretty elementary. The unfortunate part of the whole security field is that it is pretty insular. Security types write books for other security types and the rest of the world goes on its merry way doing the same old stupid stuff and getting into the same trouble despite all the handwaving that security pros are doing about the latest hack, exploit, or zero-day vulnerability.

Rather, this book is for non-computer types — the individuals who want to use computers without hassle. This book is about how to think about keeping yourself secure — the kinds of things you need to consider doing or thinking about in order to keep yourself relatively safe.

I have some bad news for you. There is no hassle-free computing on the Internet. I used to make a very bad joke that was a play on the old safe-sex saying: "the only safe compute is no compute." Inherently, for a myriad of reasons, not the least of which is the fact that we use the Internet for a whole bunch of things that it was not intended for, the Internet is simply not at all safe and it never will be.

You'd be unusual if you didn't know someone who has had their computer crash because of a virus. Perhaps it was you. You may even know someone who has had their identity stolen. While not strictly a computer problem, computers are so intimately linked to the issue that it is hard to differentiate computers and identity theft.

But, viruses and identity theft are not the only issues that are in the news. One hears increasing news of information warfare. In 2009, the federal government established the United States Cyber Command to oversee American cyber security and to protect essential elements of governmental computer and communications security[1].

One sees news reports of high military officials openly discussing the possibility of a kinetic response to information warfare attacks by foreign nations — such as China, Iran or Russia. News reports talk of repeated Chinese attempts to attack Google. Our news has been obsessed, for an extended time, with the Russian attack on the US elections in 2016. There have been large news stories about Russian attacks on European elections.

Have you heard of Stuxnet? The security community was abuzz when the Stuxnet virus made its debut. The virus, designed to attack industrial control equipment, effectively shut down the operation of an Iranian nuclear development plant. Because the virus was carefully engineered to attack a common form of industrial control computer manufactured by Siemens used in electrical, water, and gas plants, many experts were concerned that the virus would soon be released to create mayhem in cities and towns around the world. A number of experts concluded that the virus was a joint effort of the United States and Israel[2].

In the domestic arena, the news has frequently been filled with tales of hacking groups like Anonymous and Lulzsec. While Lulzsec has been effectively shut down, Anonymous has ventured into various forms of social activism. For example, Anonymous has taken credit for a number of computer break-ins at porn sites. In each of the porn site break-ins, hackers published thousands of names and credit card numbers. This form of hacking, often referred to as Hacktivism, seems to be the preferred activity of Anonymous. Before taking on porn sites, Anonymous also attacked numerous organizations in support of Wikileaks founder Julian Assange. Also, each time a hacker or group of hackers has been arrested, Anonymous has responded with internet-based attacks.

However, this is not what the typical every day home and business user of computer technology is going to be concerned about. On very fundamental level, the issue is protection of one's computer, one's business, one's privacy, and one's identity. Those issues are going to be the primary focus of this book. Perhaps the biggest issue will be the problem of viruses. Let's focus on viruses first.

Viruses are not a homogeneous thing. For that reason, many experts in the field prefer to use the more generic term of "malware." In looking at malware, we can group them in several fundamental categories.

The first category that we'll use is the term virus itself. A virus is a program code designed to do something with your computer you don't want, much like a cold virus gives you a cough, sneeze, headache and other nasty symptoms. The symptoms on a computer can be anything from a nasty message being sent to someone or to a hidden program that steals your identity, and ultimately, your money.

One thing making the discussion of these issues difficult is the term virus is also an overarching term that can be used to refer to several different subcategories of nasty programs. One of the first is a "Trojan Horse" or more simply just a "Trojan" — which makes things a little dicey when discussing the issue with individuals who are sophomoric in their attitudes and humor.

Trojan Horse comes from the legendary Greek saga of the Illiad. The Greeks destroying the city of Troy hid in a wooden horse. So, when a damaging program lies hidden in another apparently safe program, security specialists have long had the habit of describing the damaging program as a Trojan Horse. The problem with Trojans is they bring a whole range of damaging programs into the operating sphere of a computer. Some of the nasty consequences of Trojans include key-loggers and rootkits.

Key-loggers specialize in lurking in the memory of computers and recording the key strokes of computer users. The consequence of having a computer system infected with a key-logger is these programs also specialize in sending recorded information, such as passwords, to others over the internet. Consequently, Trojans are often the vehicle by which identities are stolen.

Rootkits represent another evil promulgated by Trojans. Rootkits are designed to get root or administrative privileges on the computer systems they infect. Because they get such high privileges, they are really hard to detect. In fact, the administrative level of access to the computer frequently means that the alien program is very hard to detect by the security programs which are supposed to defend against them.

Part of the problem with rootkits comes from the fact that they are designed to corrupt the hardware-controlling operating systems such as Windows, Linux, or Mac O/S. Corrupting the operating system, the rootkit has the perfect camouflage to disguise itself from anti-virus software that would otherwise detect and remove it.

Rootkits have had a very checkered career. Some companies have used rootkit technology to hamper attempts by computer users to illegally copy entertainment media. In 2005, Sony-BMG installed rootkit software on the music CDs they employed for the distribution of music. The idea behind the concept was to provide extended copy protection schemes for Sony CDs. Problematically, the rootkit installed by Sony on Windows based computers also had the effect of opening additional security holes for other viruses, or malware as they are often called.

It's not surprising to find out that there was a massive scandal arising from Sony's actions. If anything, because of Sony's actions (and others), rootkits are even more loathed, if possible, than other forms of malware.[3]

Worms are sometimes described as another type of virus. However, many security specialists regard them as a separate type of malware. At first look, worms seem to be less harmful than Trojans. However, it's safe to say that the damage done by worms is more insidious than a virus. A virus may be designed to erase or corrupt certain types of files, or it may be designed to physically damage an entire hard drive. In contrast, a worm simply replicates itself. Back in the day, when we didn't have networks, this wasn't that big of a problem. Having to rely on sneaker-net and floppy disks to be carried between computers, programs that only replicated themselves were not a big deal.

Now, we live in a networked age. If it doesn't network, we hardly consider it to be a computer. Worms specializing in copying themselves across networks, within networks, and filling up every space inside the victim machine rapidly become a problem. What makes worms more problematic is the damage is not often abrupt. The user experiences a gradual deterioration of the function of the computer. Gradually, it slows. Programs gradually cease functioning. At some point, the suffering machine becomes unusable.

The real problem from worm technology comes when a worm is combined with other technologies such as viruses or Trojans. Imagine a program designed to install a rootkit that uses network technology to move quickly over the Internet or within a local area network. Lots of damage can be done very quickly.

One type of virus I find very scary is the polymorphic or many-form virus. While most viruses have a standard set of actions that they engage in, such as what they attack or where they reside, the polymorph can change itself to hide in different places or do different things. This makes them very hard for anti-virus programs to detect and remove.

Polymorphs are evil. I've had to fight them. On a network I once ran, a polymorph got through on a system in which anti-virus had been improperly installed. The polymorph combined Trojan and worm technology, and it was designed to attack our database server. Anytime we got it off an infected computer, it migrated somewhere else, and it soon reinfected the database server. We were only able to get rid of it by taking the entire network down and cleansing every computer before bringing the network back. Going to

250 computers with two techs and two very part-time volunteers was a time consuming and painful process.

One very nasty consequence of viral infection can cause a computer to become a zombie or a "bot." These kinds of infections cause the computer to do things that are beyond the control of the user such as sending out masses of spam email or sending out masses of virus packages to other unsuspecting users. Many unethical tech types will create viruses designed to implant a zombie virus on a whole collection of computers. The result of this effort, when successful, will be something called a "bot net" or robotic network. A "bot net" can be used in unethical advertising campaigns by sending out millions of unwanted email messages. Another application of "bot nets" resides in using hundreds or thousands of computers to work cooperatively to try to flood a web server with requests for information, thereby effectively denying access to the website by legitimate users. This kind of attack is referred to as a distributed denial of service attack (DDOS).

1.1 Cookies and Sundry Varieties of Spyware

Most computer users have heard of cookies. These little bits of text, designed to help enhance the efficiency of webpage interactions, are often used to track the behavior of users on the Internet. Many times, when users initially gain access to a website, the webpage will ask the user information about their preferences. These preferences are usually stored on the user's computer via cookies.

One sees the impact of cookies every day. When surfing multiple websites, it's pretty common to go to pages such as Amazon.com and have them offer you products based on the things you have been surfing at other locations. Personally, despite the fact I've been associated with computers for years and have been familiar with the idea of this kind of tracking, I still find it mildly unnerving.

At the time of this writing, the Federal Trade Commission has been working on a proposed "Do Not Track" agreement with major web companies such as Google, Yahoo, and others. A major Do Not Track bill was passed in 2011, and several Internet tracking bills have appeared in Californian and in European Union legislature. Some are concerned that web companies will continue to track but will interpret the agreement as meaning that they will not directly market as a result of the cookies they collect.

One of the big problems in proposals made by government organizations such as the FTC is that they almost always have to reconcile the needs of different stakeholders, not unlike many systems design projects. In trying to minimize tracking and other forms of intrusion, businesses are often dependent on the information that comes from tracking to try to customize products for consumers. So, limiting tracking is a two-edged sword that cuts both for and against customers. In balancing these interests, loop holes are almost always incorporated that will permit companies to continue tracking practices.

Additionally, there are special purpose tracking cookies related to particular types of applications. One such special purpose applications tracking cookie is something called a locally shared object or LSO. LSOs are sometimes referred to as flash cookies and are associated with Adobe Flash. LSOs are very similar to regular web cookies, and like regular web cookies, they are often used to track user options, but because of their effectiveness in tracking users, there are real concerns about these devices violating fundamental browser security and providing more information about user surfing habits than users may want.[4]

Ironically, for various reasons, including Apple's unwillingness to support Adobe Flash, Flash is now considered to be an obsolete technology. Despite the moribund nature of the originating technology, LSOs continue to exist throughout the web, and they continue to put the privacy of users at risk.

Other tools exist to violate online privacy of users surfing the internet. Many unethical Internet marketers will install programs on computers that will actively send information back to the originator tracking where the user surfs and what kinds of information is otherwise held by the computer. Very similar to viruses, these programs behave like viruses and really are viruses. The generic term for these programs is spyware.

So, spyware is a serious business and one of the major things that should be done by decent security software is to try to prevent the installation of this spyware on your computer. It is not unusual to have developers expend lots of efforts to come up with news ways to install spyware on computer systems. Consequently, as with viruses, one of my first recommendations to keep spyware off your system is to keep your computer patched and to keep it protected with current anti-virus.

1.2 Social Engineering, Phishing, and Pharming

The most successful hackers in the world count on a few facts about human beings in order to ply their trade. For instance, people want to be helpful. Frequently, hackers will take advantage of that propensity to be nice by pretending to be a computer technician or some other trusted party who should have access to an organization's technology. Before anyone is aware of what is happening, you're another fly under the microscope providing data for increasing numbers of marketers.

The tactics of social engineers are varied and numerous. One common technique is to send a soliciting email. One of the more notorious styles goes like this: "Hello, I'm a political exile hoping to collect an inheritance of $20 million. Can you help me...." Either based on helping impulses or greed, people fall easy prey to the blandishments of the unethical. In this example, sometimes called phishing, the user is persuaded to provide a bank account into which the vast fortune can be deposited. Unfortunately, no fortune exists and what was in the account disappears.

In addition to basic phishing attacks, there is something called a "spear-phishing" attack. Spear-phishing relies on gathering intelligence in other quarters and using other kinds of resources to help shape the effectiveness of an attempt at eliciting information. Low technology resources like printed city directories can provide a wealth of information to the would-be spear-phishing practitioner. The concept relates to the idea that a social engineer pretends to be someone or something that they are not and they gather extra credibility by employing the intelligence gathered someplace else. By playing on the human impulse that only a "legitimate" person could know the information, the social engineer insinuates themselves further into the target's trust. So, the result is access is granted that should not be granted or a password gets revealed that should not be revealed. Information is given that should not be given. A system gets hacked.

It's interesting to note that these tactics are employed to devastating effect at all levels — against private citizens and against governments. Spear-phishing techniques were employed by China against American and British tech firms such as Google. The attacks were so successful that British military officials made ugly comments about a possible "kinetic" response.

On the calmer side, it should be noted that some writers have been stating the "kinetic" response threshold should be much higher for cyber-attacks than for other for other types of provocation. Given the fact that cyber-attacks, from governments and from non-governments, will increase, it's nice that there are some voices of reason out there. However, we should note that governments regard cyber-attacks as simply another element among the tools that they have to respond to attacks and confrontation from other countries.[5]

So, in addition to playing on the basic human instinct, social engineers play upon the fact that massive amounts of information about us are ever-present.

One of the first places they look for passwords is in the public information available about every individual....

1.3 Hackers, Crackers, Hacktivists, and Others. . .

When one thinks of security violations, one often thinks of hackers. "Hacking" engenders lots of strong feelings. An old definition of the word referred to individuals who came up with elegant solutions to complex computer problems. A "hack" was a particularly brilliant solution to a problem. With the development of widespread network computing, some of the individuals who dedicated themselves to learning about computer systems began dedicating themselves to exploring computer systems through network connections. Hence, "hacker" started to be associated with individuals who "explored" computer systems, sometimes inappropriately.

Those familiar with the origin of "hacker," concerned about the increasing negative associations, coined "cracker." In this terminology, "crackers" are individuals who break into computer systems with the idea of doing harm, of stealing secrets, or of destroying data. In contrast, "hackers" only break into computers with the idea of obtaining more knowledge.

This view of the two types is not an excessive romanticizing of the issue. Early network explorers often did look around computing systems with the idea of simply gaining more knowledge. There are innumerable examples of them gaining access to a computer system and letting the owners know about the vulnerabilities so that they could be fixed.

Unfortunately, there are also innumerable examples of the activities of "crackers." As I write this, news was released last evening that a major credit card processing firm had its systems compromised by crackers with the resultant loss of ten million credit card numbers. In a system that is fundamentally based on the faith of its users, this is quite a substantial blow.

When you start talking about hacktivists and hacktivism, the room becomes almost instantaneously divided. Hacktivism is the use of computer technology in the support of a political or social viewpoint. The tools used are numerous and varied.

One common tool used by hacktivists is to "enlist" large numbers of computers into a "bot net," as discussed previously, and have them working together to send multiple "syn," or synchronize, commands to a target server. The way in which communications takes place between computers requires that the target respond with an "ack" or acknowledgement signal. Get enough computers saying hello to the target, the result will be that the target computer spends all its time in a state of continual greeting. No other programs are able to run on the computers being overrun by hello requests. As a result, the websites being hosted by the target become inaccessible. The website crashes.

The attack just described is called a DDOS, or a distributed denial of service attack. Implementing a web of computers in different ways, through having your computer unknowingly converted into a "zombie" as part of a bot net sending hello requests or by actively using attack tools provided by some shadowy hacktivist group, denial of service attacks have proven an effective way of attacking organizations with which you disagree. Numerous examples exist where different organizations have vented their displeasure using these kinds of tools.

One example for which there are many instances involves a hacking group called Anonymous. One of their more famous adventures came with a Julian Assange incident. His news organization, Wikileaks, having received sensitive US Government documents concerning American diplomatic efforts, was pursued through the legal system by elements of the Federal Government. Having acceded to American demands that they stop supporting Wikileaks, Paypal and Visa both shut down Wikileaks' donation accounts. Both Paypal and Visa experienced substantial denial of service attacks. There were slowdowns and some actual complete denials of service.

Another example of hacktivism comes with an organization that has since been disbanded — a group called Lulzsec. Lulzsec translates from its original net-speak jargon to the following phrase: laugh out loud security. On the web, LOL means laugh out loud. LOLs means laughing out loud repeatedly. LOLs gets recreated as LULZ.

Lulzsec's biggest brush with fame came as a result of a case involving copyright violations. Sony had sued a hacker by the name of George Hotz (hacker name Geohot) over the fact he had released information about being able to alter the Sony Playstation 3 so that one could install a different operating system (for controlling the hardware of the gaming device) called Linux. Linux is an open source (statements that make up the program made openly available for modification) operating system beloved by the hacking community.

Sony decided to sue George Hotz over provisions in the Digital Millennium Copyright Act which states that any attempt to modify the content of copyrighted material is illegal. As a consequence of their actions against George Hotz, the Sony gaming network was repeatedly hacked and thousands of credit cards owned by gaming customers were stolen. Sony's gaming network was down for weeks. Lulzsec claimed responsibility for having broken into Sony's network.

As you can see, hacking and hacktivism are very serious issues, and hackers and crackers both have some very powerful tools to gain access to networks and computer systems. Many of these tools are legitimate security tools that can be used by network administrators to secure their networks, but, like most tools, these weapons are two-edged and can be turned to illegitimate purposes as well.

Some of the more famous tools used for negative purposes include NMAP, Metasploit, Backtrack, and John the Ripper. I'll provide a brief review of some. I'll also briefly examine a category of tools that can be used to capture your information in transit — a class of tools called sniffers.

NMAP is the Network MAPper. Since the most important part of any hacking attack is information gathering, gathering information about the structure and resources available on a network is a very critical issue. That is where NMAP shines. A partial list of the things that can be gathered by NMAP include the name of the computer, the operating system, any services or programs that are running on the computer, and any open pathways or ports into the network. NMAP is an intelligence gathering device. When the information that can be gathered by NMAP is combined with other sources of information publicly available about a company, its officers or individual workers, the portrait can be disturbingly complete and can very distinctly support a hacking attack against the company.

Metasploit is a system that allows security specialists to probe for vulnerabilities to determine if specific attacks will work against a computer system. Backtrack is a collection of utilities, including NMAP, that is similar to Metasploit.

John the Ripper is an example of a password cracker. While most systems keep passwords in an encrypted or scrambled format, password crackers can be used to "descramble" the encrypted information to allow unethical people with computer skills to find out what your passwords are.

Sniffers are a class of tools used by hackers and security pros that I find particularly frightening. They intercept networked data, or packets, to allow the user to see what someone is sending over a network connection. Most of the time, the odds are with you that someone won't be between you and your destination with a sniffer, but I don't care to risk it. That's why, if a site asks for data from me over the web, I absolutely insist on an encrypted connection before doing any sending. Look for the HTTPS at the beginning of the URL

(Universal Resource Locator), the internet address. We'll talk more about the importance of encryption in a later chapter.

There are lots of different sniffers available to intercept your information. Some of the more powerful include Wire Shark, Snort, and Air Snort. Wire Shark is the most widely used, but Air Snort is particularly effective in allowing someone to intercept your data while on a wireless network and Air Snort allows for the compromise of WEP wireless encryption. Again, we'll talk more about this in a later discussion of encryption.

What the collection of tools that have just been surveyed point to is the fact that there is a wide range of very powerful methods available for security professionals, hackers, and hacktivists to use in order to probe or attack systems. This means that users and businesses need to be even more careful in doing what they can to use technology wisely and in the most secure fashion possible.

Rather than being afraid of a bunch of ill-defined threats, users need to realize that there are specific actions that can be taken to minimize risks. Computer technology and the internet are a metaphor for real life. If a sufficiently bad person really wants to hurt you, they can. But, if you take the appropriate precautions, you can make it unprofitable to hurt you and you can reduce the chances of being hurt. The odds are pretty high that you will be passed by. Most people wanting to penetrate computer systems are pretty opportunistic — they go after the low hanging fruit.

The tools used by hackers and security pros point out one fundamental fact: one of the most important things that you can do to protect yourself and to limit unsavory poking around in your computer is to have a firewall. Firewalls are the electronic equivalent of the barrier that protects the passengers in a car from being harmed by the heat and mechanical processes going on in the engine of an automobile. Firewalls are absolutely necessary to limit the ability of unfriendly types to use tools to probe your computer to find out what juicy tidbits might be there.

In addition to having a firewall, you must absolutely have current, repeat current, anti-virus software on your computer. Firewalls and anti-virus will get more extensive coverage in a later chapter.

In the next chapter, we'll begin looking at how your mind, how you think, is your best tool to help secure you, your information, and your computers.

Chapter Notes

1. "United States Cyber Command," Wikipedia, accessed on 2 December 2022, http://en.wikiped ia.org/wiki/United_States_Cyber_Command.
2. "Israeli Test on Worm Called Crucial in Iran Nuclear Delay," New York Times, accessed on 2 December 2022, http://www.nytimes.com/2011/01/16/world/middleeast/16stuxnet.html?pag ewanted=all.
3. "Sony BMG copy protection rootkit scandal," Wikipedia, accessed 2 December 2022, http://en .wikipedia.org/wiki/Sony_BMG_copy_protection_rootkit_scandal.
4. "Local Shared Object," Wikipedia, 2 accessed December 2022, http://en.wikipedia.org/wiki/Lo cal_shared_object
5. "UK Warns That Aggressive Cyberattack Could Trigger Kinetic Response," Kevin Townsend, Security Week, accessed 2 December 2022, https://www.securityweek.com/uk-warns-aggressi ve-cyberattack-could-trigger-kinetic-response.

2

Your Mind, the Best Security Instrument

One of my favorite authors, Robert Heinlein, often misunderstood and perhaps loved for the wrong reasons, was often quoted as saying, "There ain't no such thing as a free lunch." Fans have often quoted that saying through the acronym, "TANSTAAFL." What the science fiction grand-master was trying to say was this: if it looks too good to be true, it probably is.

What's fascinating about computers and the Internet is that many put that very elementary piece of skepticism aside. Flush with a brand-new toy and the seemingly limitless nature of the Internet, many get caught up in the moment and forget even rudimentary rationality. "I found this really cool thing on the 'net," is one expression I hear variants of every day. Heck, I've been guilty of it myself. We get overwhelmed by the excitement of the new shiny we've found, and in the process, we lose judgment. "Look at all the cool apps I've been able to download," comes another common expression, usually not too long after unwrapping the new desktop, laptop, or tablet. When one asks what the discoverer expects to do with all the trinkets, a sheepish grin and a shrug often come back with a "I dunno, but they look fun."

In addition to the problem of the shiny, we have other issues biting at us. The difficulty is that there does appear to be free things available for the taking on the Internet. Really bright people have invented things such as open source software. Others have invented things such as the General Public License (GPL). In both instances, they appear to be (and are) genuinely free. Open source software are programs that have been written with the idea of being able to be shared with others. The source statements in a high-level language such as C++ or Java have been made available for others to look at and to modify if they want. Taking this idea even further, General Public Licenses make source code available for users to modify (if they have the background to manage it), but software licensed under this rule requires that any changes made to it must also be made freely available to others to see or to perhaps modify even further. There is a very strong belief among fans of open source software and General Public Licenses that software are ideas and ideas should be free.

The ideas of open source and General Public Licenses harken back to an earlier era in computing — one not haunted by patent trolls and other similar evil entities stalking the darkened hallways of our legal system looking for some innocent to cross a legal bridge. In the naivety of an earlier era, software was an idea and you could not patent ideas. In reaction to the growing profitability of the exploding computer software industry in the 1980s, the concepts of open source and GPL were invented because really smart people, like Richard Stallman, who invented the GPL, wanted to see a continuance of that earlier era in which ideas were freely exchanged and in which software grew organically under the tender care of many hands and many eyes.

So, as a result of these beliefs, there are some really cool programs that are completely free and that seemingly cover the entire breadth of computing. For example, the popular operating system Linux, a competitor to MAC OS and Windows, is covered under the General Public License. The very powerful computer programming language Python is open source. Libre Office, a very powerful competitor to Microsoft Office, is also Open Source. So is Gimp, a graphics editor.

These programs are great. They have many features (sometimes more than their commercial competitors) and they provide a very affordable alternative to mainstream commercial programs that can cost you hundreds of dollars or even more.

One difficulty extending from these programs is that they start to nibble away at the hard edges of skepticism that computer users need to maintain in order to remain secure on the Internet. This appears to be a one of the places where the slippery slope argument holds up. "If those cool programs are free on the Internet, why isn't everything else?" Indeed, why aren't they?

Well, there are plenty of reasons, and those reasons are going to be part of the meat of this chapter. Before, going any further, I want to emphasize one fundamental thing: there are lots of people making money on the Internet by playing on the fuzzing of the line between the truly free, such as Libre Office, and the things that really aren't so free and that may end up making you pay a price that you don't want to pay.

2.1 The Internet Wealth Formula

For those with the time, patience, energy, people skills, and tech know how, there is a pretty straight forward approach to making a very comfortable living from the Internet. Let me summarize it in the following: Offer free information, update the free information frequently, build value for users on the site, get users to sign up for your mailing list, design sales campaigns to sell them things they absolutely need to have, and start the cycle all over again. The key element to the cycle is building a mailing list. Sometimes, building a mailing list is done through building website subscriptions. Sometimes, it's done through purchasing mailing lists. In any event, for most web businesses, the absolute gold is the mailing list.

So, there is a crucial relationship between the "free" materials offered and the building of the mailing list. Most organizations who try to get you to sign up for a web subscription will offer a privacy statement or privacy policy. However, it is essential to note that there is nothing that binds these individuals to continue to adhere to the privacy policy as stated when you first signed up for the web subscription. It certainly is not unusual for the company to "amend" the privacy policy without notice. Google has done that to immense criticism. It certainly is not unheard of for web companies to sell your information to someone else. If you've never seen the phrase "preferred business partner" or a variant, then you have not been paying adequate attention. As a consequence, I've seen people innocently sign up for some web offering only to have their email address make a circuit with the result that they end up receiving some very unsavory spam.

Recommendation: In addition to your regular email from which you do your important personal and professional business, have one or more "throw-away" email accounts that you use to sign up for "free" offers. This will save your main account from being junked up with offers from the "business partners" of the website you have signed up with. Sometimes, these offers are for unsavory services or from individuals who have no compunction about sending you an attachment that has a virus or a rootkit in it.

Several web organizations specialize in providing these throw away email accounts. These organizations include, as of this writing, Guerilla Mail,[1] GetNada,[2] and Mail Drop.[3] Obviously, this is not a comprehensive list, but it is a starting place to help you find an additional resource to keep your privacy just a little bit more secure. URLs for these services are listed at the end of this chapter.

If you are a businessman, I'd strongly recommend that you have an internet use policy. In that policy, you should discourage company Internet users from using organizational email addresses to sign up for free offers on the Internet. While any organization having an Internet presence is guaranteed to be on the receiving-end of advertisements and attacks, a more general suggestion would be to do everything that you can to minimize your organization's footprint or how noticeable your organization is to those entities that might be predatory.

Recommendation: Despite the passage of the CAN-Spam act of a few years ago that was supposed to regulate bulk emailing and spam, I remain unconvinced that the legislation is effective. What the legislation was supposed to do is give the recipients the ability to opt out from receiving further spam messages. The fundamental problem is that replying to spam messages in any form acts as a kind of confirmation that the email address is a live one with real people on the receiving end. Programs exist for the random generation of email addresses for blanketing the known universe with spam messages. The net effect of any form of acknowledgement is to increase the value of the email address substantially. The consequence of this value increase is the address can now be sold as a validated email address. Despite the intent of the law, I'm not convinced that this kind of blanket sending and validating of email messages has not been accelerated by the CAN-Spam Act. So, I would suggest that you maintain the habit of never responding to spam messages in any way. Not replying to messages may actually then give an updating of spam filters a chance to actually help cut down on the amount of spam that you receive.

2.2 Free Music, Free Movies, Free Land, and Fremont

I remember Napster. It's the grand-daddy of all file sharing services. When it came out in the 1990s, it was a much more naïve time. You could download as much as you wanted, but you had to provide sharable content with the other users in the network. Many felt that they had found a way of sharing music and movies without harming anyone. I had a brief flirtation with it. As time went by, however, it became readily apparent that harm was being done, that copyright laws were being violated on a large scale, and that copyright owners would have to act. Napster was sued by copyright owners, and it lost. It morphed into a pay music service for a period of time, but it eventually died.

What's notable about Napster is it lost its court battles over the issue of distributing copyrighted materials. The would-be successors to Napster are fully aware of this fact. Being aware, they operate in the gray nether-world of the Internet. One problem of this netherworld is that the rights of copyright owners and the rights and safety of users are usually not respected. Consequently, unpleasant techniques sometimes get employed against users. A common technique used by purveyors of file sharing services is to embed spyware into their file sharing client. These elements of spyware often get deeply buried into the systems on which the client is installed, making them very hard to remove. So, the outgrowth of the installation of file-sharing clients is often a gross violation of the privacy of the individual wanting free music or software. While we will have further opportunity to look at the implications of file sharing later on, it is well worth noting that this gray realm makes file sharing generally dangerous and unwholesome.

Additionally, there has been an increase in the use of "drive-by file downloads" by some web developers. Simply visiting these sites can result in unwanted data being put onto your computer. Later, we will discuss the use of script blockers that can help to shield you against this kind of privacy violation.

2.3 Surveying the Landscape and Minimizing Your Risks

The previous chapter surveyed the general landscape of security. The breadth and variety of security threats is genuinely daunting. Knowing that there are so many risks should engender a genuine skepticism and a healthy regard for the various nasty things that can happen to you and your data. Many figures have been bandied about, but most accept the estimate that more than half the computers on the Internet are infected with viruses. Given the breadth of that figure, what can a person do?

Recommendation: Stay away from the seamy side of the Internet. While this seems like prejudice and broad-brush generalization, the fly-by-night web page operators usually do not have the energy, resources, and the skill or desire to provide a safe surfing environment for the casual unwitting surfer. These sites frequently end up being hijacked in various ways with cross-site scripts being installed permitting these scripts to redirect users to try to steal their login information.

So, what is the "seamy side of the Internet?" Obviously, my seamy and your seamy may differ. This is one place where the imprecision inherent in language can be very problematical. My prejudice would include porn sites on this list. Others that I would include would be file sharing sites, sites where lots of pop-up ads appear (there are ways to handle these — more later), and generally any site where rudeness and rough language are tolerated.

Not meaning to sound like a prude (I've been known to swear like a Missouri mule skinner), the issue with language is simple. Most well-run sites used by communities will have moderators who will help to enforce standards of civility. If these standards don't exist, I strongly question whether the site has the resources to do the things necessary to otherwise make their site a safe place.

However, I should mention a qualification to the "seamy side" rule of thumb. Often charity and religious sites end up being dangerous for some of the reasons previously discussed. The lack of resources and skill can be a huge issue for sites that are built by volunteers because the owners can't or won't hire professionals. While there is nothing inherently wrong with volunteers or volunteerism, the skill levels of the people involved can vary significantly — often leaving these websites vulnerable to being exploited by criminally minded elements that are too easily found on the Internet. So, it is not unheard of to find that some of the sites that you might least expect to be infected have, in fact, been taken over by the seamy side and have become a vehicle for passing infections on to the unsuspecting through drive-by file downloads and by the presence of cross-site scripts that are designed to take the unsuspecting to hostile websites someplace else.

Recommendation: If you own a business, discourage your employees from surfing both the seamy side of the web and from surfing religious and charitable websites while working. There are a number of ways in which you can do this. If you have a large network, your networking crew will have perimeter security appliances that can "blacklist" bad sites as they are discovered. Unfortunately, blacklists are never comprehensive because new sites come into operation every day. So, there always seems to be a contest between those who would protect networks by limiting access to "bad" websites and those who would continue to multiply the opportunities for users to wander into trouble.

Additionally, for small businesses and for home users, there are programs that can limit access by individual users. These programs have varying degrees of effectiveness. One problem they have comes from the fact that users, particularly those with even moderate knowledge of their computers, can do things that will effectively circumvent locally implemented programs for controlling where one can surf.

The classic example of software designed to limit what someone can get into on the web is the service called Net Nanny. Net Nanny is still being sold and is an effective way of keeping most children away from bad sites. There are many others devoted to doing the same thing. Just do an Internet search on Net Nanny and you'll find an amazing number of search hits. For example, anti-virus software makers often also produce "parental control" software which can be used to limit what someone can see and do on the web. Both McAfee and AVG, anti-virus distributors, market software solutions for limiting access to the web.

In addition to software solutions for restricting access, hardware solutions exist also. Hardware can sometimes be more reliable because it can be harder to get access to a network router and circumvent it than it might be to limit access to an individual PC. Typing in the key word set "parental control hardware," numerous hardware solutions came up. Most of the companies that make SOHO (small office home office) networking hardware have some sort of network filtering hardware solution available.

Beyond the issue of making sure that your kids are safe while they surf using their laptop or home computer, many young people find that the greatest portion of their interaction on the web happens through their smartphones. There are numerous "apps" will help parents limit what children can see and do on the Internet. Doing a search on parental control software for smartphones reveals numerous alternatives for protective smartphone software. When I did my search, numerous alternatives popped up, including reviews of some of the best smartphone protective applications, including sites such as Tom's Guides. I've generally had good luck with the kind of information coming from Tom's Guides.

Recommendation: If you own a business, talk with your employees. Make them aware of the implications of bad surfing habits. Remind them that their surfing leaves traces behind on the computers that they use. Their surfing may leave remnants behind with which they may not be comfortable. If you have multiple users, develop and publish an Acceptable Use Policy. I may be naïve, but I believe that most employees want to be cooperative. Give them that credit. Organizations should invest in security awareness training for employees. It pays off.

I would suggest the same kind of openness for your family. Adolescent boys are well aware of some of the unseemly things that are available from the Internet. A lot of this information will come from their buddies. Don't let the nonsense that can come from friends be the only perspective that they get regarding the garbage that can be found on the Internet. Talk with them openly about the issues and what can happen to the family computer if they go some of these places.

Recommendation: Don't let children surf unsupervised. While most will want to help by not going to inappropriate places, some will fall prey to the temptations of forbidden fruit. Realize that it is not just surfing that is a problem, but there are things such as chat rooms and social media that you need to be concerned about. Bullies and predators can really do significant harm to young children and to not-so-young children. Over the last few years, it is unusual to go more than a few days without hearing some tragic story involving children and/or adults, who have been harmed, where some have died, because of cyber-bullying. And since there are predators in the cybernetic forest who will try to manipulate people in some very frightening ways, being generally aware of what your children are doing is a good thing to help them avoid the online predators.

Other issues are involved. Letting children surf for extended periods by themselves in the isolation of their rooms allows them to substitute computers and the Internet for contact with humans. This substitution by itself makes them more prone to predators and more likely to make bad decisions regarding sites to use and software to download.

Recommendation: Have a computer separate from those used by children for handling business and professional activities. Children don't always recognize the inherent dangers posed by downloading and installing free software. You may want to consider having a special login for them separate from the primary administrative account that does not have the ability to install software. Like all of us, finding something shiny, given the opportunity, they'll download and install things that can be harmful to a computer system. Even if something is not overtly harmful, the additional overhead caused by the installation of unknown and unevaluated programs consumes space on your computer and some can end up resident in memory sucking space and machine cycles. On the bad side of things, these shiny things often do include viruses, rootkits and other evils. Sometimes they'll inadvertently download and install bogus anti-virus programs that parade around looking like security software but are actually attempts to scare you into buying their creator's dubious services. In some instances, you'll have to employ advanced removal tech to exorcise these "security programs" from your computer.

Recommendation: As stated with regard to children surfing the net, you shouldn't surf the Internet with an administrative account (an account that will allow you to install software). In fact, nobody should surf with these kinds of accounts. If you have multiple users of a computer, create multiple accounts — one for each user, including yourself. Have everyone surf with those accounts. One of the main issues for most virus writers is they design viruses to try to take advantage of administrative privileges to extend the harm that the viruses do. Inherently, surfing without administrative privileges minimizes this possibility.

If you own a business, try not to let most employees have administrative privileges. Sometimes that is just not possible. If you must let them have administrative privileges, get them to use a non-administrative account to do their surfing with.

Recommendation: Be skeptical. Be skeptical. Be skeptical. This is going to be one of the recurring themes of this book. A relatively new form of virus attack is something that I would describe as extortion-ware. This little trick occurs during surfing when you suddenly get a pop-up that declares that you have child-porn on your computer or that your computer is infected with a virus or that your files have been taken hostage. The assertion that your computer is infected is certainly correct. Sometimes a pop-up will then ask you to click on a link to disinfect your computer. However, instead of removing the virus, the click activates a script that launches the download of a more sophisticated attack on your computer. In the process, the virus will often attack your genuine anti-virus software and will block your ability to go get updates that might otherwise fix the security flaws that made your computer open to attack.

Extortion-ware, which, at the time of this writing is very commonly called Ransomware, has gotten more vicious and more sophisticated over the years. Extortion-ware has as its hallmark an unabashed attempt to get you to pay a ransom. Early versions were marked by an attempt you get you to buy their product in order to get rid of a virus. They have now evolved into pretty sophisticated blackmail with infected computers ending up with all their user files being encrypted and held hostage until one pays a requested ransom.

These kinds of programs have gotten so common that purveyors of cyber-security news now commonly have a separate news category that is devoted to keeping track of the latest variants of the malware devoted to taking people's systems hostage. Some pretty major attacks have been based around this class of software. In fact, one major outbreak took out several British hospitals.

2.4 The Inherent Complexity of Software

One of the major problems of software is the fact that it is inherently very complex. This is one of the major bases of the security risks that come from computer systems.

Many of us only glimpse the truly overwhelming nature of the complexity of software. Some have argued that software development is the most complex thing that human beings do. Whether it is or not, it genuinely is hard, as reflected by fact that so many flaws are routinely discovered in common programs such as office suites and operating systems such as Windows.

The frequency of flaw discovery has become an element of common discussion and sometimes common hilarity. Denizens who live in the Microsoft ecology are very familiar with "Patch Tuesday" and the problems created by fixing problems with Microsoft programs.

However, that being noted, my recommendation for the home user is to make sure that you make sure that your system gets the latest patches downloaded and installed. If your system gripes about Microsoft Update not functioning properly, pay strict attention to it and make sure that you find out why you have a problem. The problems that IT managers have often relates to issues that you, the home user or the small-office user, will never encounter — issues created by the need to integrate large-scale systems.

Malware developers almost always know about the flaws that Patch Tuesday is designed to address. Often security researchers inform Microsoft (and other software publishers — it isn't always Microsoft's fault) a relatively short time before the fix is developed when they themselves find the flaw in software or when they discover a virus in the wild that exploits the security flaw. Many researchers, looking to make a name for themselves, will publish the software security flaws they discover. In any event, when the patch gets published, malware developers will know about the problem. So, you'll have a remarkably short time to get your computer protected before it is vulnerable to attacks from malware.

The point? Patch your computer as soon as you find an issue. Make sure that your computer is set to download updates automatically. Apply them as soon as you can. In general, over the years, when I've had to fight major network disruptions caused by a virus or worm, it has almost always been through a computer that, for whatever reason, was not kept updated.

I've got to reiterate a disclaimer at this point. With security, there are never any guarantees. For the individual who would presume that they are completely protected because they have done everything right, they are living with an illusion. Security is all about mitigating risks with the whole concept leading to the simple issue of reducing the chance that you will be hurt. Looking both ways before you cross a street will not guarantee that you won't be run over, but it will certainly lessen the chance that you'll end up in traction, or worse....

To emphasize that there are no guarantees, a form of virus exists that generally scares the heck out of security professionals and network administrators. This is what we would describe as the "zero day attack".

Zero day attacks occur when a hacker discovers a flaw in a program, does not tell anyone about it, writes a program to explore the flaw, and releases the program into the wild (the Internet). Some of these programs, depending on how they are written, can spread like wildfire. The consequence of the spread can be that hundreds, thousands or tens of thousands of organizations can have their computing facilities, their software, their hardware, and their data suddenly compromised by the sudden appearance of a previously unknown piece of software that burns through the unlucky organizations and can cost hundreds of millions of dollars. This is one of many nightmare scenarios that can be faced by network administrators and other technical support types.

The good news is that complete zero day attacks are relatively rare. Most security researchers are extremely ethical and most study software for benign reasons (employer requests, sheer love of knowledge, and so on). But, there are those individuals who are not ethical. However, in most instances, more than one individual discovers the software flaw at the same time and the possible impact of the person with bad intent is balanced by more than one person with good intent. It doesn't always happen this way, but more often than not, it does. So, zero day attacks, while not unheard of, are generally pretty noteworthy, and they generally get pretty extensive coverage in the technical press.

2.5 The Security Implications of Social Media

Social media has become an important issue for individuals hoping to keep their computer systems secure. This is in addition to the privacy implications of social media — which we'll discuss in much more detail later on.

Social media is a broad range of applications. In looking at all of them, one must remember one fundamental fact: you are the product — not the service that is provided as an after-thought. This will have some interesting implications for privacy.

How many social media applications are there? It's very nearly impossible to count. New ones are created every day. Some of the more common ones include Facebook, Myspace, Youtube, Pinterest, Quora, and Linkedin. Numerous others exist, many for special purposes such as creating ways for scientists to be able to exchange information, creating ways for folks to share slide presentations, and enabling people to share photographs. This is not a complete list.

Focusing in on the flavor of the moment, let's talk about Facebook. One thing that every user of Facebook should be aware of is the fact that Facebook does not write all the games that are available for its users. The developers who write these applications often do not write these programs with security in mind. Consequently, these games and applications are often the vehicle for viruses to enter the systems of unwitting users.

Getting viruses through Facebook is a fairly common occurrence. So, if you use Facebook, please make sure that you have your anti-virus in-place and up-to-date. Additionally, I accept none of the game and applications request I receive through Facebook. In addition to the fact that so many of these programs want access to your personal information and they want to post to Facebook on your behalf, few are developed by Facebook itself and Facebook has been notoriously slack in overseeing the quality of these programs. In fact, Facebook has often discriminated against these applications. In a 10 October 2010 article that was entitled, "Facebook acknowledges privacy issue with third-party applications," the article states, "That activity, which violates Facebook's guidelines, raises the question whether the world's most popular social networking service has adequate systems to oversee the activity of third-party applications."[4]

2.6 Context is Everything...

One of the things I hate the most is scareware. I characterize scareware as extortion-ware's not-quite-so-vicious younger brother. With extortion-ware, when you get a notice, you are already infected. With scareware, it's a website trying to infect you. You see scareware when something appears while you surf and it pretends to be legitimate anti-virus asking you to click to download something that will make your computer secure. At this stage, scareware is relatively harmless. It's just potentially nasty content one sees on unsavory web pages. Surfing these pages, you will often see phrases like, "Your security has been compromised. Click here to remove the threat." Or, you may see a message that states that your computer has some outrageously large number of viruses (which it may have), and that you need to click on a button to remove the problems. In any event, the odds are pretty high that this is an attempt to get you to follow a link to download a program which, in turn, happens to be a virus.

In any event, it is almost certain that the program you just downloaded doesn't remove any virus. It will then often begin to act like ransom-ware by trying to induce you to purchase their "premium solution" that is guaranteed to remove threats. That usually doesn't work either.

Often, these programs, having been downloaded and installed by your innocent act of trying to protect your computer, have some very interesting properties. One property is many of them will attack your real anti-virus, preventing you from doing a scan of your computer to try to remove the virus you've just accidentally installed. What's even worse is that these programs are often designed to prevent you from getting the systems updates designed to help remove the "little helper" that's just been installed.

So, what are your options? First, try not to get these infections. I know that's more than a little snarky, but it's a good place to start. Before you click on the link, you should ask yourself, "Does this look like my anti-virus?" Unfortunately, when surfing the net, most of us tend to click impatiently on a link and when it doesn't immediately respond, we click again…and again. Alternatively, we don't look at the pop-up and we click. Even worse, frightened by the threat to our computer system, we click on the link and we put ourselves right where the malware developer wants us: downloading the "anti-virus" that's actually a virus.

Second, what if you do get the virus? Unfortunately, your options are often limited. Because the downloaded program, once installed, often disables anti-virus and your ability to go to anti-virus and update sites, a direct web solution may not be possible. In those situations where your defenses have been crippled, you may have to use a USB drive to download a virus remover from another uninfected computer.

So, what anti-virus do I use to remove the virus that has just gotten through my viral defenses? The answer is not so simple. There are many different solutions out there that have different strengths and different operating characteristics. Without repeating the details of anti-virus software which will be discussed in detail in another chapter, I'll simply say this: some anti-virus programs are better detectors and some are better removers. Get a good remover.

However, I've got to issue a warning at this point. Getting a good remover is no guarantee that you will be successful in getting rid of the scareware. Some of the programmers who implement these programs are quite clever, and I've seen some of these that deeply embed themselves into the operating system. In that case, my recommendation would be to take your computer to a tech specialist to have them supervise the removal.

Unfortunately, the tech may end up recommending that you have them burn the computer down, after saving your important files, and then reinstall the operating system. The nasty aspect of the situation is this: it's a simple economic transaction. Do you want to want to invest many hours of time thereby creating a charge into the hundreds of dollars under the supposition that you have gotten rid of all remnants of the virus or do you want to cut your losses? There are tricks that a tech-type can employ, including trying different anti-virus software or doing painful edits of the computer's registry (where all the programs and devices that the operating system can work with are registered). Please let me offer another disclaimer/warning at this time: unless you are very technically accomplished, please avoid doing a registry edit yourself. Deleting the wrong information from the registry can turn your computer into a brick.

So, given the consequences that have been outlined, you can see the basis of my earlier smart-aleck request that you try not to get these infections in the first place. Please make sure that you read the messages that come up from your computer. Please make sure that you verify that security messages that come up do indeed come from your anti-virus.

What do you do if you stumble across an attempt to send you some scareware? If it is a pop-up from a web surfing session, close the pop-up and do not pass go and do not collect $200. Immediately close the pop-up.

When I'm feeling paranoid, I do a control-alt-delete to immediately go into Task Manager so that I can immediately close the application without even touching the offending pop-up window. Control-alt-delete is often referred to as the "three-fingered Windows salute" — a not so oblique reference to the traditional obscene gesture. This command sequence activates a primary Windows control menu allowing you to shut your computer down or launch a program called Task Manager. One of the things that I would recommend that you consider doing is doing a search on "Task Manager Tutorial" and read one of the numerous lessons that are available on the subject. There is a pretty good one available at one of my favorite blogs, "Bleeping Computer." If the preceding information seemed too challenging, you can always contact a tech-savvy person for assistance. However, getting help from your cousin's really-smart 13-year-old who's really good with computers is not always the wisest action.

The Internet continues to insinuate itself into all aspects of our lives. It provides us with many different kinds of entertainment. As noted, many of us live on social media. Also, the Internet is essential to our work. We use computers as a nanny for children (often not a very good thing). It has long ago supplanted discreetly wrapped mailed magazines as the most effective method for transporting porn. People are now talking about the arrival of virtual robotic prostitutes as a possible cause for a precipitous drop in human population. We now control our homes remotely from the web. We monitor our children and pets from the web. The list goes on.

So, it would seem that this segue into a discussion of dating sites would be a letdown from the previous paragraph.

You've seen the ads: continual blandishments to take the guesswork away and to use technology to find your absolutely simpatico life-partner. The problem with these sites is they are notoriously bad in keeping user information secure. In most instances, putting one's personal information onto a dating site has had the same effect as writing one's most intimate personal data on a sheet, making a gazillion copies, and handing them out to random strangers on the street with the implicit hope that none of them will stalk you or try to steal your identity.

Recommendation: A dating site is a hazard to your security. Even though it may sound arrogant for me to say so, consider trying to meet someone in one of the "traditional" ways. You can go to church. You can audit a college course. You can join a club. You can volunteer. All of these would afford more ability to control what goes out about you than what is routinely distributed by some dating sites.

What are some of the problems of these dating sites? One problem is that many cameras will embed GPS information about where a picture is taken inside digital data of the photograph. Quite naturally, this can give serious clues as to where a person lives. Beyond that issue, dating sites have been known to not secure the web site programs themselves — thereby allowing users to see all kinds of underlying personal information captured by the site ostensibly to provide the data necessary to be able to find the ideal match. The problem is that more information than one would want shared can get shared with individuals one might never want to ever interact with.[5]

Let's take it a step further. Suppose you don't want to meet your soul-mate. Suppose you just want to "hook-up" with someone. Without passing judgement on that, I would remind everyone that putting sensitive information on the web is just not a good practice. It isn't smart, for example, to store credit card data on a website — any website. What happens if the site's payment system gets hacked? Despite the inconvenience, it is wiser to re-enter payment information each time that you need to do so. Putting information on the web about your very personal interests is even more risky that storing payment information. The odds are very high that the information will, at some point, get compromised. Do you want that data plastered all over the web? Well, it happened to participants in the notorious hook-up site Ashley Madison.

2.7 Change your Business Practices...

In security book after security book, security experts note that employees remain the largest source of security breaches and data loss for organizations. The reasons for this are numerous: employees falling prone to social engineering attacks, employees engaging in stupid internet surfing habits, employees playing with social media during work hours (and introducing a virus to your network), employees downloading music and picture files that are infected and illegal, employees having bad passwords that can be easily guessed, and employees deliberately stealing money and/or compromising critical organizational information. We won't examine this list comprehensively at present, because some of these topics deserve their own chapter and others have already been discussed.

2.8 Social Engineering Attacks...

Social engineering is based on the fundamental ways in which human beings are wired and is consequently very hard to defeat. Without going into the details that will be addressed in a later chapter, human beings want to be nice and social engineering plays upon that propensity. As an example, a man shows up pretending to be a job seeker who has lost his resume. Desperate, he asks to have a receptionist insert his jump drive so that he can print his resume out again. The jump drive is inserted, and carrying a virus, it infects the business network. Here's a place where training and consciousness raising will pay businesses back immensely.

While training users on the issue of social engineering, one should also remind users that they should not use passwords that are easy to guess or that can be easily figured out from information commonly available about all of us. Many forget that lots of information is available about all of us in old-fashioned technologies such as city directories. In addition, massive amounts of information are available through growing numbers of Internet-based background check services. One should remind users that they should not use common passwords such as spouse names, children's names, pet names, and so on. All of this information, very commonly available, will be the first things tried by social engineers in an attempt to hack into a system.

2.9 Social Media at Work...

Some of the dangers of social media have already been discussed. They are a ready path for viruses to enter a computer and a network. Personally, I'd recommend that businesses block Facebook, My Space, LinkedIn, and Pinterest at work. We'll talk about other implications, including personal and organizational reputation, later on.

2.10 Business Considerations of Music, Movie, and Picture Downloads...

In addition to what has been previously discussed regarding the general unwholesomeness of shared movie, music, and software downloads, businesses should consider another implication for illegal downloads. This practice puts the organization at extreme risk for severe civil and legal liabilities. One should note the fact that organizations supporting the rights of copyright holders, such as the RIAA (Recording Industry Association of America), have been quite aggressive in prosecuting copyright violations against both individuals and businesses. If a business is found to have someone participating in illegal file sharing, they are not terribly sympathetic to a claim by the business that the downloading was not authorized. The outgrowth of this set of circumstances is that a business can end up with potentially fatal liabilities.

Here's where one business truism really applies. Unless you are a tiny "mom and pop" business, you really need an information technology use policy that addresses issues such as illegal file-sharing. In addition to providing guidance and giving the basis for actions against repeated violators, a written set of policies can help to demonstrate your diligence in trying to uphold legal behavior.

2.11 Inherent Risks from Employees and How to Ameliorate Them...

Judging whether or not someone is going to be an honest employee is tough. There's plenty of theory in human resources manuals about trying to effectively hire the best person. That's beyond the scope of this book. However, there are some things that one can do to help lessen the impact of potentially dishonest workers. In examining these issues, it's worthwhile remembering that the majority of security violations in business come from the employees of the business, and there are things that can be done to limit the impact of employee dishonesty and stupidity.

Don't kid yourself that you are making things more efficient by employing so few employees that you and they can't afford to take the time for a vacation. If an employee is doing something dishonest, then it becomes harder to detect if they are the ones exclusively doing that job. Besides, the time off helps to mend outlooks and attitudes, thereby making it less likely that a person will do something foolish.

Business-folk should make it a habit to employ division of labor. Having one person responsible for all aspects of a task takes away the inherent accuracy check coming from multiple eyes. Additionally, one person, acting in isolation has more opportunity to do something dishonest. For example, having the same person process your accounts receivable and make your bank deposit is a prescription for possible disaster.

2.12 Be Authentic

I admit it. This is a silly introduction to a serious subject. However, one of the most important issues that both individuals and businesses have regarding technology is proving that you are who you say you are. Tech types refer to this as the process of authentication. You may be doing that on your own system. You may be doing that on systems for your business. You are more than likely doing that for more than one account over the Internet.

The process of authentication can involve multiple different things. Most of us think of it as consisting of a username and a password. Security specialists will tell you that other things can be involved also. These other things can include something you have or something that is intimately a part of you (something you are).

The tools of authentication ultimately all have some real drawbacks. User names can be inferred — especially when they know your name and when everyone in the organization has the same basic format for the user names. Passwords can be often be guessed also. Here is one area in which people really do not effectively use their wits. Too often, they will use passwords that have some personal or emotional significance. They will use children's names. They will use spouse names. They will use pet names. Anyone, like hackers, doing basic social engineering, as discussed previously, can gather these data items and they will use them as one of the first attacks against you or your systems.

Furthermore, passwords can often be inferred by simply watching you type. The technique in social engineering is called shoulder surfing.

Other tools in authentication can also be a problem. Consider something you have. The classic example of this is the bank card. You have to have one to access ATMs. However, bank cards, like other tokens that represent you as you, can be stolen. So, they have problems also.

Let's consider biometrics or using your own body to prove you are who you say you are. We have an image of biometrics that comes to us from the movies. A retina scan gets done and you have instant verification of the identity of the user as a result. Alternatively, movies will often depict the hero swiping a fingerprint and having the person immediately verified and authenticated.

The images that we have of biometrics from movies and popular literature have some fundamental problems. They don't account for the fact that, in the real world, illnesses and other physical problems can interfere with the accuracy of the verification. Spy novels with multi-factor authentication never mention that James Bond may be bloated from a round of drinking the previous night. This can affect the shape of his hand or even the accuracy of his finger print ridges. The images from movies certainly don't account for the fact that enrolling a person is frequently hard and intrusive. Those of us who have been fitted for contact lenses know how uncomfortable getting a retinal scan can be. So, biometrics is often highly inaccurate and often difficult to use. Most security specialists will wrinkle their nose at the idea of using biometrics by themselves.

We can list as many forms of authentication as you want, and there will almost always be some serious drawbacks to each form. So, what do we do? The answer is multi-factor authentication. Use the tools in combination to increase the security and accuracy of authentication. So, we might use a credit card in combination with a pin number to allow you to get access to your money with your bank card. We might use retina scans with a security card to allow you to get access to a government defense installation. With multi-factor authentication, you are limiting the impact of the flaw of each different form of authentication you are using when combining different forms together to get access to the service, programs, or facilities you want to access.

If the service or website you use provides multi-factor authentication as an option, my general recommendation is to activate it. However, there are exceptions to every rule, and we'll outline those exceptions in the following paragraphs.

The unfortunate fact is that pretty much any good thing can be abused. Multi-factor authentication is no different. While I strongly encourage the use of multi-factor authentication, when it is available, I would be very skeptical of its use in all situations. Some social media sites that specialize in selling your personal information to advertisers have been found to take presented second factor data and use it to more effectively track you and to sell that information as an additional tidbit that can be pedaled in order make money.

For example, on some social media sites, some people have uploaded contact lists. When you provide your phone as an element of multi-factor authentication, that datum gets combined with information from friends who have uploaded detailed contact information to allow the social media company be able to more specifically target you.

Recommendation: By all means, use multi-factor authentication, but consider the context of where you are using it. If financial transactions are involved, it is a no-brainer regarding the use of multi-factor authentication. Multi-factor authentication increases your security immensely. However, I would suggest that its use with things such as social media, especially when social media sites have a history of dedicated data-mining and when they can use the extra tidbit to more effectively follow you, may be something that you want to think twice about.

2.13 Backups

Media, like people, dies. No matter how good your hard drive, your SSD, or whatever form of storage that you use, your storage will eventually die. Most of us understand this depressing fact, but we almost always choose not to think about it.

So, pushing the inevitability of media death to the backs of our minds, we continue to toss data items onto the desk top or onto our documents folder, under the assumption that what was will continue to be. And then, our drive fails, leaving us with a disastrous loss of time, effort, and information.

Writing this, I can hear so many sympathetic nods and so many impatient takes where someone sub-vocalizes, "tell me something that I don't know." Further along the path of that thought sequence is the observation that backups are tedious and they take too much time. Well, that is both true and not true. Let's talk some specifics.

It is important to remember that we are talking about cloud storage, and there are some important things about cloud storage that we all need to keep in mind as we decide how we are going to use it. Elsewhere in this document, the idea of cloud storage was investigated in greater detail. In that context, you have an important decision regarding whether or not you are going to use a free or a paid service for backup. Considering that issue, I would advocate for the use of paid services because there is less of a probability that the provider will traipse through your information to find insights allowing them to be able to more effectively sell something to you.

Almost all cloud storage systems will allow for different styles and forms of backup. Not wanting to go through the advantages of incremental versus differential versus full backups (because a tedious consideration of the different forms and their advantages is something that is important to security specialists and is probably not important to the casual user of computer tech), suffice it to say that we really do want a service that automatically captures the information that we would save, for example, under "My Documents", and automatically copies to a folder in the cloud.

This auto-duplication service is extraordinarily common in the services that provide cloud-based backup for users. The beauty of this kind of feature is that it takes the tedium away from backups and it removes the excuse that so many of us have for being lazy, for being thoughtless, and for not doing that we all know that we should do and simply back our data up, dammit.

So, if you don't have a cloud based back up service, the magic question is this: why don't you?

There are a number of alternatives that you can choose from. At the time of this writing some of the better regarding cloud backup services include: IDrive, Backblaze, Acronis, pCloud, and Carbonite. They will differ based on a number of factors, including: cost and features. Simply make sure that the service will meet your needs where the most important need is the automatic capture of information that we sloppily toss someplace.[6,7]

Not only do these services protect the biggest part of the expense in your use of computer technology (your time!), but they can also have really super important work-flow benefits in that they will allow you to work pretty much anywhere and still have access to the latest version of your efforts.

2.14 Conclusion

This chapter focuses on how important your thinking is to guarding your security on the Internet. We looked at a wide range of issues such as GPL software possibly enhancing your security, social media having some really seriously negative impacts on security, signing up for free services and providing your email address having a bad impact on security and privacy, music services potentially impacting you in a bad way, online dating exposing more information than you might like, and authentication and identification impacting how safe your information is. Finally, the importance of getting past hang-ups regarding backup was discussed. Modern cloud backup services relieve a lot of the tedium of backing up.

This list of issues helps to highlight that a lot of security starts and ends in how you think about things. A sense of skepticism about what you are presented is a starting place, but realizing that there are tools that can help you secure yourself such as cloud backup services and multi-factor authentication will go an additional step to making your use of computing technology safer and more reliable.

Chapter Notes

1. "Guerrilla Mail – Disposable Temporary E-Mail Address," Guerrillamail.com, accessed December 2, 2022, https://www.guerrillamail.com/
2. "Disposable Temporary Email," GetNada.com, accessed 2 December 2022, http://getnada.com/
3. "Save your inbox from spam," Maildrop.cc, accessed 2 December 2022, http://maildrop.cc/
4. "Facebook Acknowledges Privacy Issue with Third Party Applications," Los Angeles Times Blogs, accessed 2 December 2022, http://latimesblogs.latimes.com/technology/2010/1 0/facebook-acknowledges-privacy-issue-with-third-party-applications.html
5. "Security of Internet dating sites examined," United Press International, accessed 2 December 2022, http://www.upi.com/Science_News/2012/01/13/Security-of-Internet-dating-sites-exa mined/UPI-15061326504004/
6. "Best Cloud Backup Services 2022: Free & Cheap Backup Storage," Aleksander Hougen, Cloudwards, accessed 22 November 2022, https://www.cloudwards.net/award/best-onli ne-backup-services/
7. "The 5 Best Online Backup Services of 2022," Albert Bassili, How-To Geek, accessed 22 November 2022, https://www.howtogeek.com/790664/best-online-backup-service/

3

Essential Security Tools

There are a set of essential security tools that should be in the "digital tool box" of every individual using computing devices. By saying this, we can note that while we might have the idea of the computer foremost in our minds, we are talking much more broadly, and we should include smartphones, tablets, and pretty much anything else even remotely recognizable as a computer in this list.

The list of essential security tools that will be recommended here include: anti-virus, firewalls, password managers, VPNs, and encryption tools. We'll discuss each one of these tools in this chapter, but because some of these concepts are covered in more detail in other parts of the book, the treatment of those items, like VPNs and password managers, will be done rather lightly in this chapter.

3.1 Anti-virus

Let's consider anti-virus first. As I type this, I can figuratively hear the whooshing of a goodly portion of the audience turning their nose up at the idea of anti-virus. The usual accompanying explanatory exclamations fit into the general categories of, "well, I have a Mac..." or "well, I have a Linux box." Fair enough. Both Macs and Linux boxes are more resistant to viruses and most forms of malware than are Windows-based computers. However, let me be explicit at this point: they are not immune. Let me repeat. They. Are. Not. Immune.

Let's take Macs first. The various versions of Mac OS X that are out there are all based on a version of Unix, BSD Unix (Berkley Systems Development).

The various versions of iOS, Apple's mobile operating system for iPhones and whatnot, are based on Mac OS X.

The beauty of the Mac is that its operating system is based on the Unix Kernel. The Unix Kernel is constructed in such a way as to make it very difficult for malware to be able to attack a computer and its functionality without the active cooperation of the Kernel, which malware is not going to get. So, as with Linux, Mac based systems continue, even to this day, to be far more resistant to viruses than are Windows-based computers.

Still, nonetheless, apparently foolish companies, including Norton, Bitdefender, McAfee and Eset, continue to persist in pedaling anti-virus for Mac. The magic question at this point is this: why? The answer: because Macs are not immune to viruses — despite what Apple would have you believe.

So, the question then becomes, "should I put anti-virus on my Mac?" That's an easy question that has a really complex answer — one that goes to the heart of computer and network security theory. The first part of the answer to the question comes from how comfortable you are with risk. Do you engage in any risky behaviors? Do you surf seamy sites? Do you download lots of "free" applications like games (such as employing plants to murder zombies)? Do you open lots of attachments from friends without looking at them or even expecting them?

The next part of the answer is this: what do you use your computer for? Are you running a business off it? Is your financial life ensconced on this bit of electronics? How important is the computer to you and to your life?

You see where I am going with this. Bad behavior combined with a high level of importance means that the answer is going to be a resounding yes — put that anti-virus on your Mac.

At the time of writing, some of the different manufacturers of anti-virus for the Mac include: Intego, Clam, Clario, McAfee, Norton, Avast, Bitdefender, Eset, Trend Micro, Avira, Kaspersky, Sophos, Malwarebytes, MacKeeper, and F-Secure. This is not a comprehensive list.

I have had experience with a number of companies and have generally had good experiences with McAfee, Avast, Bitdefender, Eset, Trend Micro, Sophos, Malwarebytes, and F-Secure.

Let me emphasize, Macs do get viruses. At the time of writing, the Silver Sparrow malware was detected on even very recent generation Macs running the M1 processor. Hundreds of thousands of Mac users, according to Macworld, ultimately got infected with the code. There are several reasons. Increased use of Macs mean that they are more likely to be targeted by the underworld developers of malware.[1]

3.2 Anti-virus on Linux

What was said about the security for Macs applies even more strongly to Linux based systems. For malware, Linux, to employ a tired old saying, is a hard nut to crack. Again, the center of the power of Linux as something resistant to viruses is the Linux kernel.

Much of what was said about Macs regarding the issue of whether or not you should have Linux antivirus also applies to Linux. How much risk can you tolerate? How bad are your behaviors?

There is very little that should be stated with absolutes when regarding security and computing, but here is one: if you use a Linux-based computer in a mission-critical role, such as that of a file-server, then you absolutely should have that computer protected by antivirus. You would be foolish not to have that protection.

If, on the other hand, your Linux computer is simply something that you use for your day-to-day computing chores, you can certainly take a lighter view of the issue of anti-virus.

There are a number of manufacturers of Linux anti-virus. An organization called Ubuntu Pit identifies the "Top 15 Best Linux Antivirus Programs." These include solutions from organizations such as Sophos, Comodo, ClamAV, F-Prot, and others.[2]

Recommendation: If you are using a Mac or a Linux computer, do a risk assessment. How do I use my computer? If you use it in a risky fashion or if the computer is the basis of mission-critical work, protect it with antivirus. Linux-based servers should be protected. Otherwise, if your computer is used for just personal work, you can probably skip anti-virus on both of these platforms. However, with the very rapid growth of Mac-based computing and the fact that this increase in popularity makes them a much more attractive target for potential hackers, I would be less than at-ease in leaving my Mac unprotected.

3.3 General Issues Regarding Antivirus

At this point, it is probably important to understand some basic issues regarding antivirus. The first thing to understand is how antivirus works — which consists of two levels. The two levels include heuristics and definitions. Let's start with the second issue, definitions, first. Definitions are specific bits of information that antivirus programs use to halt viruses. These specific bits of information are based on the idea of the footprint of a specific piece of malware. The footprint of a bit of malware are the specific behaviors exhibited by the malware in performing its nefarious task. Where does it reside? What does it attack? How does it come into a system? This is the footprint of a virus.

Antivirus definitions will tell the antivirus program how to recognize and remove the malware in question. So, it is really super important for antivirus definitions to be kept up to date. This importance is only highlighted by the other level at which antivirus programs work: heuristics.

An easy way of explaining heuristics is to describe them as rules of thumb. One way of presenting this is use the old idea of "if it walks like a duck and it quacks like a duck, it must be a duck."

An individual looking at the idea of heuristics might be pardoned for thinking that computers could be protected by just using heuristics. Despite all the talk that we are subjected to regarding machine learning and artificial intelligence, we need to emphasize the fact that computers remain profoundly stupid. The whole idea of heuristics is that we ask computers to make judgements. Frankly, given the state that computers remain in, I wouldn't want to stake my life, my computer, my financial safety, or whatever else that I value residing on my computer to the "judgement" of software.

Here's the problem: as things continue to stand, heuristics are either way too strong or they are way too weak. Being way too strong, they will tend to label too much stuff that has nothing to do with being a virus as a virus. I had one piece of writing shareware that my antivirus hated. It continually described the program as a virus and kept trying to remove it. After struggling with this for an extended period of time, I finally gave in to the persistent security program and went to another solution for my writing utility.

On the other side, one can frequently adjust heuristics and help to train the security program that things otherwise recognized as viruses are, in fact, safe. The downside with training your software is you risk weakening your defenses too much so that other hostile threats can be let through.

So, one might be inclined to completely dismiss heuristics as part of anti-virus defense. In that vein, one might presume that the only protection one might have against viruses is the use of definitions employing the footprint of a virus to stop the activities of that virus.

This idea of relying exclusively on definitions has the problem that new viruses are constantly coming out. One of the big problems in the ecosystem of computing is there is a huge array of what many in security would be inclined to call "script kiddies" or "code monkeys." These pejorative terms are used to describe individuals who have a minimum of technical skills and who take the work of master hackers and modify it to create their own variant of malware.

You see, to be a real "hacker," you need to acquire a great deal of in-depth technical knowledge of how both hardware and software works. You really need to be an ace programmer. Script kiddies or code monkeys don't come anywhere near that level of acumen or skill. However, their minimal knowledge, combined with a virus design toolkit created by a master hacker, leverages their impact immensely, and it exponentially multiplies the impact of this subspecies of hacker by creating variants of existing viruses just different enough to make definitions potentially ineffective against new virus variants.

So, because of the existence of script kiddies, definitions are not proof enough against viral computational attacks to rely exclusively on them. You need to complement their work with heuristics.

In addition, there is something else to be worried about. You will often hear security types speak in hushed terms about "zero-day" attacks. These are where a new, previously unknown, exploit appears to attack systems from a direction that was previously unknown. For security types, this is extraordinarily scary.

The good news, however, is this. Zero-day attacks are really quite rare. They just don't happen that often, but when they do, they can wreak massive havoc in systems that are, in effect, not protected against them. The existence of zero-day attacks is another argument for the continued existence of heuristics. There is no guarantee that the heuristics of an anti-virus program will detect the zero-day exploit, but not having them built in will increase your exposure substantially.[3]

So, which programs use heuristics and which programs use definitions? The happy answer is this: pretty much all anti-virus programs use both.

Recommendation: In addition to having anti-virus on your computer, make sure that your anti-virus is up to date. Most anti-virus program providers will periodically provide new definitions to protect your system. Most anti-virus will download these new definitions regularly. Let them do their thing on your behalf.

Recommendation: Many commercial versions of anti-virus are sold on a subscription basis. Find out if that is the case on your system. Find out when your subscription expires. Don't let it expire. Period.

Recommendation: Keep your computer and all its programs updated. Manufacturers of software regularly provide updates for their programs as new vulnerabilities and flaws in the program are discovered. These updates can be referred to as patches, as service packs, or simply as updates. Whatever they are called, they are your friend. Making sure that these flaws are fixed are super-important to your security.

Guess what? Pretty much every type of computer system is dependent on regular updates.

One of the biggest vulnerabilities on most systems comes from Java — a program embedded in your browser that allows you to run more advanced programs that are embedded in web pages. Without delving into the theory of how Java is supposed to work, it is designed to create a protected work area for these advanced programs to work safely. Well, it doesn't always work that way. Java is attacked a lot more than it should be.

Recommendation: Don't just skip by the common request that your system makes to update Java. Update it!!

3.4 Know your Enemy

The famous quote from the ancient Chinese general is often summarized as follows: "If you know the enemy and know yourself, you need not fear the result of a hundred battles." At this point in our concern about the use of anti-virus, it is probably worth a brief diversion to consider some of the different sources of malware. One source we have already partially considered: script kiddies. In addition to our reflection on script kiddies, it probably is worthwhile considering other sources: criminals (organized or not) and state actors (governments).

Criminals have discovered the wonderful world of attacking computers, systems, and people as a way of making a profit. They work alone or as part of organized crime groups. In either event, the effect is the same. It increases the hazards for people using computing technology.

One of the nastiest outcomes of the criminalization of hacking is something called ransomware or extortion-ware. Ransomware is a virus variant that once it infects your computer, encrypts the entire contents of your computer and demands payment to give you the key to release the encrypted contents of your computer system.

The existence of nasty creatures such as ransomware makes anti-virus and keeping your computer up to date absolutely essential. It is heartbreaking to see your collected work, residing on a computer, now completely inaccessible because it is encrypted.

Do not let yourself be diverted by something that may look similar. One might call this category of potential attacks on you and your computer as scareware. The grand hallmark of this creature occurs when one gets a lurid message on your screen that screams, "your computer is infected!" They may then follow up with a message demanding that you click on a link to clean your computer.

Recommendation: Above all, never click on those links. By clicking on them, you may be potentially downloading a virus instead of fixing your computer. Often, good anti-virus will screen those pages out, but if it happens that it doesn't, don't be a partner in allowing you and your equipment and software to be attacked. Don't click on the link.

3.5 State Actors

Let's talk about the last major category of potential attackers: state-actors. We are talking countries: the United States, Russia, Korea, and Iran — as starters. However, I'm sure that the British, the French, and the Germans have activities in this realm as well.

State actors are the authors of some of the nastiest malware that can come your way. There are many examples of viruses and other like creatures that are made by state actors. A great example of this was the Stuxnet virus that many believe was created in a joint effort between the United States and Israel.

State actors are often the best sources of the dreaded zero-day exploit — exploits that haven't been seen before and for which there are no anti-virus definitions. Remember zero-day exploits are relatively rare, and security experts tend to note their appearance. Stuxnet, designed to attack Iran's nuclear production infrastructure, had an unprecedent four Windows zero-day exploits built into it. This fact about Stuxnet has been reported by a number of sources, including the online magazine, CNET. The term often used for viruses made by state actors is advanced persistent threat.

State actors have been behind some very generally nasty forms of malware. This includes some frightening ransomware, and it also includes something that I find particularly frightening: key-loggers, or programs that are designed to capture, save, and forward the keystrokes that you type. You can see the danger: someone who installs a key-logger on our devices has a way of uncovering our passwords.

While state actors tend to try to attack significant sites impacting large industrial, economic, or military targets, their impact affects TCMITS (the celebrated man in the street). In addition to impacting significant businesses, the attacks of state actors increasingly bleed into the domain of the average guy or business.

The purpose of this reflection on state actors is simple. It is part of a reminder that you really should have anti-virus on your computer systems, and hope that the heuristics are able to find and filter potential malware coming from state-actors.

3.6 A Primer on Different Types of Malware

In the spirit of knowing your enemy, we really need to put ourselves on the same page regarding the different kinds of malware that you are likely to encounter when (notice the choice of word) you encounter malware. This will help you make the right choices in fighting the problem. Additionally, if you take your computer to a professional, being able to accurately describe what is happening while using the right terminology is likely to save you time and aggravation in dealing with whatever computer-type you enlist in the repair or rescue of your device.

For purposes of this discussion, we'll focus on the following types of malware: viruses, trojan horses, polymorphic viruses, worms, root kits, and macro-viruses. Security types would not be happy with this selection because we are actually presenting some categories that actually overlap, but this choice of categories is probably the best to give non-professionals the perspective they need in order to more effectively protect themselves. We will consider each category in turn.

3.7 Viruses

The term virus has been used so broadly as to lose a large part of its effective meaning. From these standpoints, many in the field prefer a deliberately broad term, malware, to encapsulate all the different possible variants of things that can attack your computer.

A major manufacture of free antivirus, AVG, offers the following definition of virus:

"One of the oldest types of computer threats, viruses, are nasty bits of malware that hijack your computer's resources to replicate, spread, and cause all sorts of chaos."[4]

Fussy technical types would offer the additional qualifier that the program in question needs to be a separately identifiable entity. Why that particular qualifier is important will become apparent soon.

So, viruses are programs that have the ability to move about and many have the capability of causing damage, sometimes severe, to you or to your system. Frankly, this damage can include some distinctly disastrous physical damage to your computer.

3.8 Worms

Of all the different kinds of malware that are out there, the pure worm is probably the least directly damaging of all the different variants of malware that we can encounter. In its purest manifestation, worms do nothing more than replicate, and they most usually replicate over networks. The biggest problem with worms is that they will replicate into whatever space is available. This can mean that individual systems or even networks can cease to function properly under the increasing burden of worms taking up more and more space, more and more memory, and more and more processing time. So, worms are nothing to laugh at.

Worms are classically a creature of networks. Network engineers will frequently describe how worms propagate via network shares — or shared folders that can be accessed via network connections. You are infected with a worm. You access a shared folder. The shared folder now has a worm, and that worm will likely be replicated onto the next computer whose user accesses the shared network folder. And so, the fun continues.

Please note this disclaimer: some viruses have characteristics of multiple malware categories. It is not unusual to find a worm that also has the characteristics of a virus in the modern sense of a program that can directly attack the hardware and programs of the computer on which it lodges. One nasty variety might be a worm-virus that propagates among multiple users with the idea of lurking in each computer in order to log key strokes. Nasty.

Rootkits are a particularly bad type of virus designed to persist in a computer even after the computer has been rebooted. One way of permitting a virus to persist between sessions is to put the virus onto the part of the disk that controls the startup or booting of the computer. Startup portions of the disk are called the boot sector and the master boot record. Another way to implement a rootkit is to have it hide in the commands that guide the start-up operations of your computer. These programs are built from a partially changeable section of the computers permanent memory that used to be called BIOS (basic input output system), but that is now called UEFI (unified extensible firmware interface).

So, when tools like anti-virus do their work, they remove the traces of the virus they can find, only to have the computer get re-infected the next time that the computer starts up. The effect of this situation is that rootkits often only can be removed by completely erasing the operating system and reinstalling it from scratch — erasing anything that may be on the computer.

Macro-viruses are a form of malware that are created from the operation of applications. There are two issues at play here. The first is something called a macro, which is a way of recording commonly used command sequences in tools like Excel, Word, Powerpoint, or Access. The second are scripting languages embedded in applications.

Macros can be used to automate common complex operations for advanced users. Unfortunately, they can also be used maliciously to attack the computer and their information. Surprisingly complex operations can be created through this simple process of recording keystroke sequences. Really nasty viruses can be written using this seemingly simple facility for recording keystrokes.

An infamous instance of a macro-virus was something called the "I love you virus." This virus spread through emails, and it would read email contact lists to use that information to spread itself. Having gotten infected, the primary impact was that you would have to suddenly explain professions of eternal love to many on your contact list. I got infected, and I had some uncomfortable times having to explain myself as a result.

However, there is another level of danger residing in many applications because it is not uncommon to include a scripting language for building even more complex commands. Microsoft includes something called VBA (or Visual Basic for Applications) in its Office programs. What this means is that even more complex viruses can be constructed and embedded in data files associated with Office programs.

What is the net impact of this issue? First, it highlights the overriding importance of having anti-virus tools. Second, it also illustrates some significant changes that are necessary for the behavior of computer users who are more likely to protect themselves.

Recommendation: Be very cautious of opening files, particularly Office files, sent to you via emails and the web. They can and do often contain viruses. If you are not expecting someone to send you a file, don't open the file. Never, ever, ever skip past the warning provided by your email client warning you about the dangers of opening attached files. Read them and consider them. They can seriously and nastily attack you, your data, and your privacy.

The examples provided are not serious examples of how truly vicious a macro virus can be. For the most part, these were relatively silly examples. They can be a lot nastier than those provided in these examples.

3.9 Trojan Horses

This type of malware is the prime example of why you don't want to get that free copy of Photoshop from your buddy. Too often, the unethical people who remove the copy protection or digital rights management (DRM) from software will include an extra gift in the program files that, as they say, "keeps on giving." It has happened, more than once, that they also hide code into the program files that does enabling unpleasant things to the system on which the software is installed. Like the Trojan Horse represented in the stories of Homer, the hidden logic of the virus acts just like the hidden Greek soldiers in the horse and it does something that the installer did not bargain for.

It does not have to be direct maliciousness of the person who breaks the DRM of the software getting copied. Too often, they are as careless with their own systems as they are with the copyright of software that they illegally copy. The files and programs that they distribute gets infected through their carelessness.

Recommendation: Stay away from illegally copied commercial software. First, if you run into technical difficulties with the software, who can you go to? The questions that you run into may or may not be answered in user forums. Too often, the really significant questions just aren't answered or they are just enough different as to not be applicable to your situation.

Recommendation: In obtaining software, try to use software that comes from software stores associated with your systems. These software stores work really nicely for Macs, multiple versions of Linux, and for Windows. There are some pretty nice Android stores also. However, in my experience, the Mac stores are probably the best run and most reliable.

Recommendation: Be cautious with freeware, shareware, and General Public License software. Give preference to software associated with organizations such as GNU, the Free Software Foundation, Libre Office, Open Office, and MySQL. There are others.

If they provide a digital summary of the data files making up the program, make sure that that you download it. You should compare the digital summary, or hash, to the hash provided by your downloading program. This will help to ensure that something extra has not been injected into the program file that you just downloaded. It is an additional protection against trojan horse logic having been inserted into the program that you just downloaded.

3.10 Polymorphic Viruses

Look, the odds are pretty high that you are not going to encounter a polymorphic virus. I have. It left scars, and I am going to do a brief discussion of this kind of entity as a result.

Polymorphs scare the hell out of me. They go against pretty much everything that a person working in technology thinks that they know about viruses. They are often new enough exploits that heuristics will not have been adjusted in order to detect them. Furthermore, because polymorphs will change what they do, footprinting will be initially difficult.

The time I had to fight a polymorph was because a sloppy tech released computers without the requisite protection software on it. Consequently, systems got infected. Because the virus was a polymorphic virus-worm, we'd remove the virus only to have the system get reinfected. We finally had to shut the entire network down so that we could go to every system and remove the virus from each system before we could bring the network back up. What lovely fun.

I include this description here because I wanted to let you know how truly evil polymorphs can be. This directly targeted our executive information system — pretty much limiting our ability to carry out our daily business.

3.11 Conclusion... No

So, this brings us to the end of our discussion of malware. Is this list comprehensive? No. It was formed basically by two factors: what is commonly out there and what scares me. Before leaving the issue of malware, I should directly address a fallacy possessed by numerous inexperienced computer users: if one anti-virus program will protect me, won't more than one provide me with additional protection? The answer is no. Most decent anti-virus programs work fairly effectively — if you behave yourself. Adding additional anti-virus simply adds to the amount of work that your computer has to do without a commensurate increase in effectiveness. It slows your computer down without doing any better on finding viruses.

Now, we need to note that there are other ways in which your system can be attacked, and those other ways in which you can be attacked often come pretty directly from the fact that your device is most likely to be connected to the internet. Well talk about network and internet issues next.

3.12 Firewalls

The biggest network and internet issue stems from the fact that you need a firewall. It's that simple. No argument. It is what keeps the wild manifestation of outside networks from coming in, all at once, to your computer system in order to hold a disease-fest on your device. Or it will help to keep external users from coming in and poking around your device without your permission.

The good news is this: practically all manufacturers of operating systems for devices will provide firewalls as part of the basic security configuration for what they sell. You will also find that manufacturers of anti-virus will also include firewall software as part of their basic security solution. This inclusion isn't just a nicety for the anti-virus maker. It is an essential part of trying to keep a system basically protected from external threats. Most viruses, for example, come into systems from the Internet — the basic realm that firewalls are supposed to protect you against. So, the inclusion of a firewall as part of a security suite only makes sense for security software vendors.

So, if we accept the basic idea of a firewall as a barrier between you and the external world, the question then becomes, "how does this class of software work?"

Without going into a lot of detail about how the Internet and networking functions, we'll just simply say that the data envelopes coming into your computational device, packets, are all numbered to represent different kinds of information a device can receive. These numbers, or ports, will allow different programs on your computer to receive data when information is destined for them. If a program isn't used by a device or if the wrong kind of information is sent via a port, then that packet will be screened from entering your computer. This is the action of a firewall.

There is a whole elaborate theory as to the nature and function of firewalls. That theory is beyond the scope of this document. Those who wish to transmogrify into networking nerds should probably seek information on that process someplace else.

Firewalls can be software on your appliance or they can be separate dedicated devices. For the vast majority of us, firewalls will be software on our appliances, and it will most often be integrated with the anti-virus software that is also essential in the protection of your system.

There are some standalone firewalls that are worth mentioning. Some of the better ones, at the time of this writing include:

- Zone Alarm
- Comodo Firewall
- Tiny Wall
- Glasswire
- Free Firewall

Again, this is not a comprehensive list. However, I would like to offer some comments on some of these tools.

First, Zone Alarm has been a substantial presence in the field of computer security for a very long time. They are quite experienced in the development of security tools. At the time of this writing, getting Zone Alarm may seem a bit confusing because of all the different options that are available. Not only do they have a complete security suite, but they also have a standalone firewall. Furthermore, they have both free and paid for versions of their products.

So, the magical question remains: which one should I buy? I'll share my experience. Generally speaking, I have found the free version of security packages from reputable manufactures to do a very satisfactory job. Are these manufacturers reputable? Generally, yes.

Why does anyone buy a solution? The answer in all areas of security has to do with the issue of features and configurability. If you have a network that you need to protect, you may want to consider a purchased solution — giving you more capability to craft a solution matching your needs.

For me, I have been perfectly happy with free versions of most security software.

3.13 Password Managers

We will cover password managers more extensively later on. However, I wouldn't use modern computational devices without one. They allow you to do your computing without having to rely on your memory and without having to rely on a notebook.

Why are password managers more secure than just simply having a password notebook? Password managers will allow for the automatic filling of passwords into the various sites that one goes to. That helps to protect you against nasty creatures such as key loggers. Furthermore, if someone discovers your handwritten password notebook, you are had. In contrast, password managers will encrypt your passwords and will submit them for verification in an encrypted format.

Password Managers will allow you to use different passwords for different locales. You'll no longer be tempted to reuse the same cotton-picking password in every bloody location because you have too many to remember. Security-types will tell you that reuse of passwords is a huge no-no. One password leak may compromise multiple accounts.

3.14 VPNs

VPN stands for virtual private network. They are an essential tool. The term will be used pretty broadly here, and will include technologies such as privacy-oriented browsers like TOR (The Onion Router).

As discussed in other parts of this book, Internet service providers often act against us. One way in which they do so is to sell our browsing history to "business partners." So, this means that if you have a fascination with porcelain dolls and you demonstrate that through your surfing habits, advertisers will also be aware of that fact from your ISP and you will be targeted with ads as a result.

A direct consequence of this economic transaction is that yet another element of your privacy is up for auction. Consequently, a person is really taking a risk with their security and privacy if they do not act to hide themselves and their movement of data across the Internet.

VPNs go a long way to hide you and encrypt the data that you are sending out over the internet. They can, for example, hide your IP address as well as making sure that the random person intercepting your data packets can't simply peruse your information. They do this by encrypting information on the way to its destination. This removes a lot of the invasiveness of ISPs in hawking your data and surfing habits to the highest bidder.

However, in addition to providing an additional layer of security to the data that you put out on the Internet, there are some situations in which VPNs are absolutely required. Those situations include:

- When you are connecting to do work for your business
- When you are using public wireless.

When you use a computer to connect with your business, the tech-types who enable this to happen are completely remiss if they do not insist that you have and use a VPN. Proprietary or even sensitive information can easily be intercepted as a result of the use of remote connections. Generally, most web connections use a protection method such as SSL or TLS. However, the fact of the use of remote connections at all makes a VPN essential for business operations.

Here is one of the major problems created for us by the possibility of remote interactions with a computer that may be half way around the world. There is a class of hardware/software known as a sniffer. Sniffers specialize in capturing packets going over a network — allowing the user to be able to look at the data packets being sent from one location to another. Yes, they will allow them to see everything being sent by you to and from your destination.

Sniffers have some really legitimate applications for network engineers. The problem is that many of them are free, and they can't be kept out of the hands of individuals with dubious ethics.

I could not have done my job as a network engineer without tools like Wireshark, Snort, or Kismet. All of them will allow you to intercept and inspect data packets. At the time of this writing they are all available for free downloading.

Recommendation: Never use public wireless without a VPN. Look, you are dependent on the safety and security of the business' network. For a lot of the private networks that I have seen in business, that alone gives me the willies. Additionally, you are also trusting the other people connected to that public wireless. How can you be sure that the mild-mannered guy in the corner doesn't have Wireshark on his computer?

Some of the tools available for capturing packets will allow you to view what is going on with another person's screen in real time. Being curious, I've played with these tools. If you have to use public wireless, which I do not recommend, please use a VPN to keep technically capable looky-loos out of your business.

Living in a rural area, it is a common thing to see "mom and dad" coming into a Barnes and Noble to use the publicly available wireless to order Christmas presents from Amazon for the grandchildren. Don't be those people.

Recommendation: Invest in a decent quality VPN for use in your daily computing. Free VPNs are available, but they are generally worth what you pay for them. Too many of these free solutions will plaster your screen with advertising banners or will severely restrict your bandwidth for connecting with your remote location. They are particularly famous for doing this if you have been liberal in your consumption of bandwidth.

Who makes a good VPN? There are a number that you can choose from. I would recommend that you look up the VPN reviews from PC Magazine and from Tom's Hardware. They regularly review VPN services and they have some excellent insights on the quality of the products.

3.14.1 VPN-like Tools

In this category, we'll put tools like The Onion Router (TOR). There are a whole bunch of other privacy browsers, but in the service of some degree of brevity, we'll focus on TOR.

TOR is VPN-like in several ways. First, it creates an encrypted connection for you. Second, it will hide your IP address from your destination. So, they won't have as easy an access to who is surfing their site.

In some interesting ways, TOR does some stuff that is better than a VPN. It randomly routes your traffic through an array of volunteer relay sites. The consequence is that you can look like you are surfing a site from someplace like Sweden, when, in fact, you are in the United States.

TOR was created by US Naval Intelligence as a way of allowing dissent in politically restricted countries to be able to connect and communicate via the

Internet. There are two elements of bad news. The NSA has been working to compromise TOR relay nodes, and it has succeeded in compromising some. So, the TOR network is not totally proofed against snooping by the US Government. You should also be aware of the fact that the sometimes-circuitous routing of data via TOR can impact, negatively, the functionality of your interactions with a website. At least, that has been my experience.

I mention TOR to let you know that this is a generally decent alternative to having more security while on the web if you do not have a VPN. However, get the VPN.

3.15 Encryption Tools

Encryption is the process of mathematically altering the contents of data to make it not easily readable to someone who has no business knowing your information. Let's talk about some reasons why having encryption tools are absolutely essential in the modern computing environment.

In order to do so, let's start with a couple of very likely scenarios. In looking at encryption, remember you can do disk or drive encryption or you can do file encryption, where you scramble and protect just individual files. In the first of our examples, we'll talk about traveling with a laptop. In the second, we'll talk about the propensity that many of us have to store information in the cloud.

Let's look at the first scenario. Many of us travel with electronics. We may have a vacation, but most of us will still have a need to connect to the job. If you are traveling for business or for personal reasons, disk or drive encryption is absolutely essential to you.

Most laptops will come today with an encryption option. Use it. What happens if you get distracted and someone snatches your laptop away from you? Do you really want personal information — possibly personal financial information — to get into the hands of a thief? This alone justifies the extensive use of encryption software.

Recommendation: If you are a Windows user of laptops, activate Bitlocker full-disk encryption. If you are a Mac user of laptops, employ File Vault full disk encryption. If your laptop is based on Linux, you can employ something called LUKS which stands for Linux Unified Key Setup. It can do full disk encryption. In any event, don't risk your data on a mobile computer whose drive is not encrypted. To do so is extremely foolish.

If you don't want to use the tools provided for you by Microsoft and Apple, there are a lot of good alternatives, some of them free. Veracrypt is an example of a fairly well-respected open source utility that fits into this category.[5]

When security types talk about encryption, they will often talk about encryption at rest or encryption in movement. Encryption in movement is often handled for us. When we see an "https:" connection, that is encryption while your data is moved between your computer and some computer on the internet. I would never use a service requiring things like passwords that did not use "https:" connections. Period.

Being observant of whether or not your connection is http or https has a number of benefits for you. The biggest benefit coming from the fact that you are protected from things such as "man in the middle" attacks (MITM).

MITM attacks are a real issue. Remember, you generally don't have control over the path that your data takes to its destination. It can take a circuitous route through a whole bunch of different stop-off points and networks. There is no guarantee that your data won't be intercepted and examined on its way to its destination. This is what MITM attacks are all about.

If someone intercepts your data or a part of it on its way to its destination, and they alter it to their advantage, it looks like your data on the receiving end, but it isn't. On the way back, the process happens all over. It looks good to you, but it isn't.

Https connections will substantially guard against this kind of nastiness happening to you and to your data. The good news is that this type of connection is pretty much standard on the Internet and most browsers implement it by default.

Recommendation: Don't be trusting. Don't just assume that the URLs on internet connections are properly implemented. Check internet addresses when you go to sites. Look for the telltale Https. Demand it. Find another place that provides the service or the information that you want. To mix metaphors from another part of life: there will always be another site (or ship?) passing your way soon.

As we began talking about encryption, two scenarios regarding the use of this tool were mentioned. In the second, potential issues were raised regarding use of online storage or cloud storage. The use of cloud storage makes it really important to be able to encrypt individual files. Let's look at some potential problems with online storage that may cause you to want to employ file-level encryption.

Here is the problem: you need to pay attention to whether your cloud storage is free or not. If it isn't, a big question is this: how do they make money from you? Nothing is ever truly free. When an online storage service is supposedly free, it is more than likely true that you and your data are the product they use in order to make their money. More basically, the drives they use for storage cost, and they have to make that cost somewhere.

So, you get free storage from, most famously, from places like Google and Microsoft. However, numerous others also claim to offer this free service, including at the time of writing: Dropbox, Amazon, Megaupload, and many others.

Here is one of the tricks: more than one provider of free data storage has been found to go through the data in the files stored on their service to allow them to make indices of the contents of your files so they can more effectively target you for advertising.

Microsoft, sometime back, found out that Google engaged in this practice, and their advertising department "made hay" with the information, admonishing individuals to not get "get scroogled" by Google. Interestingly, Microsoft, soon after, was found to engage in the same practice.

Recommendation: If you use online cloud storage, get a utility that will allow you to do file encryption. There are several reasons for this. First, you are circumventing the propensity the providers have for waltzing through your data. Second, you are applying the fundamental principle that security experts pretty much universally recommend: defense in depth. Don't just rely on the security provided by the provider. Add your own security to your data that you put into the cloud by encrypting the files before you store them on somebody's server someplace.[6]

An objection can be raised at this point: what if I have purchased storage rather than using free storage? I should be safe, right?

Not necessarily. Some of the companies that provide paid for services have been found traipsing around the data of users. I generally would not trust them either.

Furthermore, there is the idea of defense in depth. Don't just rely on just one protection. Make the individuals who want to attack you, your data, and your privacy have to really work for it by making them go through multiple protections to get at you.

Most attackers are creatures of opportunity. Like the potential mugger who sees a tough-looking person and decides to skip the trouble, hardened systems and data will usually be passed by because they are not worth the trouble.

So, what are some of the better file-encryption programs? I would look for reviews in the popular technical press. PC Magazine and Lifehacker both have some good reviews on different file encryption programs. A couple of good file encryption program include: AxCrypt and GNUPG (GNU Privacy Guard).

GNUPG is worthy of being singled out in this discussion. It is an example of General Public License software that is not only open source and free, but modifications of the software must be made freely available as well.

The great advantage of open source and GPL software is that many eyes have generally reviewed the software to help ensure that flaws have not been incorporated into the code implementing the software. This issue of discovering flaws is particularly significant in encryption where some very notorious algorithm failures have resulted from the use of proprietary programs.

When looking at encryption, we really do need to delve, at least lightly, into some of the theory surrounding the process. There are two basic forms of encryption: public and private encryption. Let's look at both.

Private encryption is also called shared-key encryption. Sometimes, it is called symmetric encryption. Basically, it requires the use of the same key or password to encrypt or decrypt the information. This is the form of encryption that most of us are generally familiar with. It carries with it the fundamental problem of sharing or distributing the key between the individuals who wish to communicate. There are many ways in which this is done. One of the more common ways of solving this problem is to send the key for decrypting a message "out of band" or using a different method for sending the message. You might, for example, ship the encryption key to the person you want to send an encrypted message by snail-mail before emailing the encrypted message. In another alternative, you might simply hand the person the key in person before they take their leave of you. You might even send them a separately encrypted email.

On the other hand, if the only person who is involved in the communication is you, there is no problem. You might be storing information in the cloud and using encryption to prevent some nosey third party from being able to read your information. At which point, there is no key distribution problem.

However, there really is still a problem. You have to keep that encryption key safe. You cannot lose it. Most good symmetric key encryption algorithms will leave the data they encrypt so completely scrambled that if you lose the encryption key, you have lost the data.

Public encryption is an interesting alternative to private or symmetric encryption. Public key encryption will always involve a public–private encryption key pair. When you are sending a message to someone who has a public encryption key, you can use that key to encrypt the message, and only the person (hopefully the person you intend the message for) with the corresponding private key will be able to decrypt the message.

The net consequence of public encryption is that if you are using it to communicate with somebody else, you really don't have a problem of getting the key to person you want to communicate with. As described in the previous paragraph, all you need to have is their public key. Use that to encrypt, and according to the theory of encryption, the only person who will be able to decrypt the message is its intended recipient through use of their private key. In public encryption, the keys always travel in pairs: public (which everybody knows) and private which is known only the person having that key.

So, in looking for encryption programs, and I recommend that you do, pay attention to the kinds of encryption that the program will support. That will be very explicitly described in the information provided about the program.

Issues that you may want to consider will include:

- Public encryption or private encryption: your choice on this issue will depend on what you intend to do with your information.
- Whole disk encryption versus file encryption: this really isn't a choice. You will probably end up using both, but in different situations. Use disk or drive encryption to protect the complete contents of your laptop when you travel. Use file encryption to keep files safe when you put them into the cloud.

3.16 Conclusion? Really!

So, in this chapter, we have looked at a set of essential tools that a smart user of computational devices should have in their "bag of tricks" (apologies to Felix the Cat). These tools include: anti-virus, firewalls, password managers, VPNs, and encryption tools. Notably, in discussing anti-virus, the position advocated here is that you should do a risk assessment. If you are Windows user, the prevailing assessment holds: get anti-virus. If you are a Mac or Linux user, consider the level of risk before making that decision. Many will state that you don't need anti-virus if you work with Mac or Linux. I am unready to paint with such a broad brush and will stand firm with the assertion that you need to assess your level of risk.

Chapter Notes

1. "Best Mac antivirus software 2022: Security software compared," Karen Haslam, MacWorld Magazine, accessed 14 October 2022, https://www.macworld.com/article/668850/best-mac-antivirus-2022-which-av-software-should-you-install.html

2. "Top 12 Best Linux Antivirus Programs in 2022," Medhi Hasan, Unbuntupit, accessed 14 October 2022, https://www.ubuntupit.com/best-linux-antivirus-top-reviewed-compared/

3. "Stuxnet attackers used 4 Windows zero-day exploits," Ryan Naraine, ZDnet, accessed 20 October 2022, https://www.zdnet.com/article/stuxnet-attackers-used-4-windows-zero-day-exploits/

4. "What Is a Computer Virus?" AVG Signal Blog, accessed 20 October 2022, https://www.avg.com/en/signal/what-is-a-computer-virus

5. "The Best Encryption Software for 2022," Neil J. Rubenking, PC Magazine, accessed 2 November 2022, https://www.pcmag.com/picks/the-best-encryption-software

6. "Five Best File Encryption Tools," Alan Henry, Lifehacker, accessed 2 November 2022, https://lifehacker.com/five-best-file-encryption-tools-5677725

4

The Risks of File Sharing

As mentioned previously, I remember using Napster in its early days when it was a brand-new concept. At the time, file sharing seemed like a cool thing. However, there was always a quid pro quo aspect to file sharing communities. I could get access to their music, but I was also expected to share my music — to share access to my computer, and the idea of open access to my equipment was something that I was never happy about.

So, I immediately disconnected from Napster and have never seriously tried a file sharing service since then. On the other hand, I have had to clean up the messes left by individuals who, in their ignorance and selfishness, have gotten themselves in a lot of trouble by using these kinds of technologies. Being the cleanup man has continued, even to this very day.

The lesson that I got from Napster is that there is no free lunch. It may sound good and it may look wonderful, but there is always a hook, a payoff that has to go somewhere.

So, as the years have passed, I've watched the names of the file sharing services come and go. Unfortunately, every one, has had one or more serious "hooks" that in my mind make it not worth the risk. The list has been long. It has included Napster, Limewire, Frostwire, Bearshare, Pirate's Bay, Warez, Bit Torrent, and many other important ones that I am sure I am overlooking. I've even have had individuals tell me that their version of file sharing is different, that theirs is genuinely a community.

Not too long after I stepped away from the Napster community, it was sued into oblivion. Yes, I know that some mutant version of it is still in existence and is part of a pay service called Rhapsody.

The issue with the massive law suits that happened with Napster is this: owners of copyright are very serious about protecting their intellectual property. You can see this in very high resolution in the SOPA/PIPA fight that occurred between the tech community and the owners of intellectual property (read that as Hollywood). This isn't the first thing done by IP (intellectual property) owners to protect themselves. It's just another step in a rather long story that has left a rather long list of casualties behind it. The nature and tenor of these adventures is something that owners of technology should be aware of before venturing into the murky realms of modern file sharing.

One thing that individuals should know about who would participate in the free bounty of file sharing is the Digital Millennium Copyright Act or DMCA. DMCA has driven a lot of the behaviors of business. It has left a lot of casualties in its wake, and it is further evidence that our lawmakers are clueless in writing laws to set up an effective legal environment that effectively organizes computer technology and its transactions.

The provision of the DMCA that is going to most impact users is the restriction against modifying the intellectual property of copyright owners. This means that the very act of removing copy protection and other factors limiting copying, now euphemistically called DRM or digital rights management, is now illegal. Most digital media can't be illegally copied unless it is modified to permit the copying to be successful.

While this provision of the DMCA should not be problematic for the typical person who wishes to share in the bounty of what someone else has made copyable, it has coincided with a large increase in the aggressive behavior of copyright owners and their affiliated organizations. For example, discussed in a paper about the Recording Industry Association of America (RIAA) from the Electronic Frontier Foundation entitled, "RIAA vs. The People: Five Years Later," on one day in 2003, representatives of the music industry sued 261 average citizens for copyright violations.[1] Over the last few years, the RIAA has sued tens of thousands of regular citizens. In fact, the only real issue is the scale of the lawsuits. According to the website ARS Technica, the RIAA claims only half of what other observers have put forth for the number of lawsuits. Some observers put the figure for the number of lawsuits implemented by the RIAA at over 40,000.[2]

So, what is the basis of all this anger? I would suggest that it is the fundamental belief by intellectual property owners that they have been horribly wronged by technology. In an article published by Forbes magazine, the recording industry in a lawsuit against one of the manufacturers of file sharing software asserted that they were due trillions of dollars in damages. Tim Worstall, writing in the 24 May 2012 article for Forbes, puts the issue into perspective in this way: that amount of damages being claimed is more than the entire productivity for each and every citizen of the Planet Earth for an entire year. Worstall puts his skepticism in this way with his headline, "The RIAA: Do Not Believe a Word They Say, Ever, For They're Claiming $72 Trillion in Damages."[3] For additional information on the perspective that the owners of intellectual property have been overstating their case, you may want to consider the following: http://www.ted.com/talks/rob_reid_the_8_billion_ipod.html.[4] In this talk, Rob Reid, founder of the company that created the music service Rhapsody, does an exceptional and very humorous job of completely skewering what many in the tech field regard as some very large scale disingenuousness by the entertainment industry.

So, how does this squabbling impact the average user of computers in a small business or in the home? The answer is this: these people are flat-out mean. Do not risk running afoul of them. Getting a cheap download on an album is not worth the risk of the misery that these people can visit on you.

The question might then be raised of how they find out. One very common technique is rather simple. It's called a honey pot.

Honey pots have been used by network administrators as a device to protect their networks for a long time. Typically, a network engineer will set up a computer, a server, that she/he wants hackers or other explorers to examine. The basic idea is that by providing an attractive target to the explorer, the other more important computers will be ignored. In the process, the network engineer hopes to gain information about the attacker, like who they are and where they are located.

So that's what happens on the Internet. Organizations attempting to protect intellectual property will join the peer-to-peer file sharing system in the hope that somebody will come to them to get parts of the files needed to make the entire song, album, or movie. When someone stumbles upon the honey pot, the implementor has their IP (internet address), and the fun begins.

Initially, someone discovered to have participated will get a rather harshly worded cease and desist letter. If they have multiple violations, they are taken to court and sued.

Having dealt with student computing in a dormitory, I've had the "privilege" of seeing a few of those letters. We fixed the problem and the issue ended there.

However, there have been many individuals who have not been so fortunate. One poor soul left his wireless unencrypted, and one of his neighbors started surfing it (and downloading movies illegally). Not being in a position to be able to fix the situation, he received multiple editions of "the letter," and he ended up in the bowels of the legal system, having to fight the charge that he was a thief.

So, can these attempts to lure potential file-sharers, these honeypots be circumvented? The answer is yes, at least for the time being. But, I'm not going to go into the methods because with expanded surveillance of what we do being conducted across the board, I would question their over-all effectiveness. Second, after studying the technology, I remain unconvinced that the technology is ethical. Third, I remain supremely unconvinced that the potentially high risks are worth the minimal savings that one could accrue by not paying for content.

4.1 The Wonderful World of Filesharing

There are a whole bunch of different programs and services that are openly available to support participation in file sharing activities. While not intended to be anything that is anywhere comprehensive, I'd like to go over some of the better-known ones.

Let's talk one of the most notorious, Limewire, first. As of the time of this writing, when one searches for Limewire, you see a variety of often conflicting entries that pop-up. Some refer to the fact that as of 2010, the company is under court order to no longer distribute Limewire. Others simply describe it as a discontinued software program. Still others tout the idea that you can download Limewire right here. Frankly, I'd stay the hell away from any and all of those. I'd regard the ethics of the individuals trying to distribute software that is apparently dead as being pretty iffy at best. Consequently, I would be afraid that these practitioners would install spyware and viruses on my computer. This concern leads to the ideas in the following paragraphs.

Having seen a number of computers infected with programs like Limewire over the years, I've always been convinced that the primary purpose of these kinds of tools has always been to gain access to the other information stored on the computer. Limewire and the like have always been a premiere venue by which spyware, root kits, and other forms of nastiness come into a computer system

To top it off, from my experience, programs like Limewire have had the nasty tendency to try to embed themselves as deeply into a computer system as they can — making it hard to remove. I've seen technicians struggle with the removal of Limewire only to finally get it off and have the computer system consequently crash.

So, you won't be surprised when I say that if you are a small business, immediately stop doing business on a computer with Limewire or any other popular file sharing software installed. As a home user, you should not place any sensitive personal information on a computer infected by Limewire or by any other of the popular file sharing programs. Furthermore, you should not do any Internet banking on any computer so compromised.

Let me add to this by also noting that the individuals who are still out there distributing Limewire appear to be ignoring the injunction and they appear to be ignoring the intellectual property rights of the individuals responsible for its original development. So, do you think that individuals like this would have any compunction about embedding nasty spyware into their product? Would they have any compunction about stealing personal information found by the spyware?

If you have any iteration of Limewire or related software, new or old, get it off your computer, pronto. Get a tech to help you with it, but make sure that you back up your important data first, because the computer may not survive the exorcism process without needing a complete reinstall of the computer's operating system.

There are several reasons to get Limewire-like programs off your computer as quickly as you can. First, as mentioned, Limewire, at one time, embedded spyware into your computer system, and it opened major security gaps into your computer that can be exploited by other writers of malware. As far as getting free movies or music is concerned, it really needs to be weighed against the resulting increased risks.

There are a number of popular peer-to-peer file sharing systems available for users. While many do not have the gross security and privacy risks of Limewire, all of them have issues with the fact that they are often employed by end users to violate copyright law and copyright owners know it. Consequently, most, if not all, of these networks have likely been infiltrated by the owners of the intellectual property being distributed. So, use of file sharing software can lead to substantial civil liability.

Programs that can be included in the list of popular peer-to-peer file sharing systems currently includes (but is not limited to): Frostwire, GnuTella, Bearshare, and Bit Torrent. Many have made claims that one or more of these tools are superior. These claims are relatively unimportant when you consider the fact that they all suffer from the same problem that Limewire suffered from:

the potential of honeypots from intellectual property owners. In the category of "for what it is worth", the current iteration of Limewire has reinvented itself as a site that "is back as an open music and entertainmentNFT marketplace for creators and collectors."[5]

In addition to the risk of getting entangled with copyright issues, files obtained from these kinds of sources tend to have, historically, a relatively high chance of viruses being embedded in them. Since the movie and music files are often broken up into chunks and hosted on different computers, each with unknowable levels of security and virus prevention, the simple act of downloading these files exponentially increases the chance of collecting a virus. Fundamentally, your computer remains as clean as the computers that it networks with, and with these kinds of networks holding files (sometimes called streams) hosted in parts on many, many different computers, there can be no guarantees of safety. If you get into the underworld of these shady file storages, you can find information on the "quality" of different stream sources. These are still no guarantee because you are relying on the technical and evaluative capacity of the people making the assessment. Given the unknowns, this is pretty thin assurance.

I'd make the strong recommendation to simply stay away from these tools.

An overview of some of the more notorious locales where you can download stuff: One of the more notorious locales where you could download free files was Warez. Warez is both the name of a site and a cultural term in the digital underground. Warez is an example of leet-speak. You see variations of leet-speak in strange little substitutions of numbers for letters such as "h311o' for hello. But, there are other variations of leet-speak including short-hand for words, in this case "warez" for wares, which is a term for illegally copied content or software.

The process of creating "warez" often involves several steps. The first step is the removal of any copy prevention — sometimes referred to as copy protection (archaically) or in modern terms, "digital rights management" or DRM.

The second step can involve, in the case of a movie or book, breaking the file up into smaller sub-files. The third step is building a control-file which identifies the various parts of the whole file or stream. The fourth step is distributing and indexing the content.

Originally Warez referred to a series of offshore websites where illegal content was hosted away from United States copyright laws. Those sites have long since been shut down. Today, the term refers to something very much like Limewire, Frostwire, and other peer to peer networks. As such, Warez has the same advantages and disadvantages.The Pirate's Bay was started by anti-copyright activists in Sweden in the early 2000s as site for tracking and indexing bit torrents. While this site remains open and is one of the most visited sites on the internet, it has faced pretty much constant legal battles over the last few years from intellectual property owners. It's been shut down more than once and has even had its servers raided. Since it indexes bit torrents on peer-to-peer networks, the honeypot issues mentioned previously apply to this service as well.

Megaupload was a file sharing and storage service that aggressively sought out and stored large amounts of copyrighted material. They were consequently raided and shut down in a multi-nation police operation. A significant outcome of the incident is that lots of individuals who had subscribed to Megaupload as a data storage service lost access to their data. The reason why Megaupload is mentioned here is as another cautionary tale. Don't store your data on a service that aggressively flouts international copyright law.

The variety of peer-to-peer services that seems to be currently the most persistent are the ones based on Bit Torrent. Bit Torrent persists because breaking up files into chunks and making them available is a convenient way of sharing and moving large-sized files. This tool may not have the same danger of embedding itself into your system and taking information from you. However, it has every bit of the other dangers that are present with peer-to-peer file sharing. The two biggest dangers are the issues of being infected with a virus from one or more of the streams that you are getting from someone else, and the very real possibility of running afoul of copyright owners, particularly those of the entertainment industry.

The final moral of this long story is this: free isn't free. It can lead to compromising your privacy, to compromising the integrity of your computer, to loss of data stored on the Internet, and to involvement with some very unpleasant aspects of copyright law.

Chapter Notes

1. "RIAA V. The People: Five Years Later," The Electronic Frontier Foundation, accessed on 5 December 2022, https://www.eff.org/wp/riaa-v-people-five-years-later

2. "Has the RIAA sued 18000 people…or 35000?" Nate Anderson, Ars Technica, accessed on 5 December 2022, http://arstechnica.com/tech-policy/2009/07/has-the-riaa-sued-18000-people-or-35000/

3. "The RIAA: Do Not Believe a Word They Say, Ever, For They're Claiming $72 Trillion in Damages. Updated, See Correction," Tim Worstall, Forbes Magazine, http://www.forbes.com/sites/timworstall/2012/05/24/the-riaa-do-not-believe-a-word-they-say-ever-for-theyre-claiming-72-trillion-in-damages/

4. "The 8 Billion Dollar Ipod," Rob Reid, Ted Talks, http://www.ted.com/talks/rob_reid_the_8_billion_ipod.html

5. "Limewire Originals," Limewire, accessed 5 December 2022, http://www.limewire.com

5

Working to Maximize Your Privacy

Computers and the Internet have real implications for privacy, and privacy hugely overlaps security. It turns out that information gathered about us can be used to attack our security—in various different unexpected and nasty ways.

It seems like we have continual news reflecting the concerns that many have about the fact that privacy seems to be disappearing faster each and every day. Every day reports come in about systems that have been hacked, exposing credit card and personal information for people who previously thought themselves protected. Identity theft and protection have become major industries. The basic process of gathering information about your target is an essential element of the process of hacking.

In contrast, many are concerned about the role of government in computing and communications. Government has worked to monitor many aspects of our lives. The consequence of these two factors means that privacy is at risk. Two primary questions flow from this risk. What are the primary sources of privacy loss? How does one minimize their exposure to privacy loss?

5.1 The Rise of Big Data

It used to be relatively difficult to find information about someone. One pathway was to research public records. To do that, you'd have to go to government offices and ask for access to the public records. Alternatively, you could go to the public library and look through the city directory to find tidbits such as the target's address, employer, marital status, age, and so on.

In this world, stealing identities required work. Information was scattered and not always easily accessible.

The advent of the electronic age changed that. For example, the Federal Government started to computerize the storage of information from the Census. It made great sense. Being able to retrieve and manipulate information more efficiently was a huge help for politicians, bureaucrats, and decision-makers. This is just one situation in which large amounts of information were placed into databases describing people.

However, there was a dark side to this increased efficiency. The Federal Government routinely made census data, identifiable to individuals, available to private businesses. But, despite the fact that information was going to business, the real problems really weren't yet manifesting. The computing facilities maintained in this era were primarily isolated monolithic mainframes that could not be easily accessed by others than the authorized members of the business.

The development of the public Internet changed all that. Cookies! Business rapidly discovered they could track customer behavior much more effectively by using cookies to find out where web surfers have been and what interests them. Additionally, many businesses found it profitable to provide discount cards to customers in exchange for detailed customer information extracted from each transaction that businesses then store in increasingly detailed and complex databases. Furthermore, business went increasingly online. Amazon and a host of other businesses made incredible amounts of money by going "on the web."

The consequence of all this was that personal data, once isolated on corporate mainframes, grew exponentially and became available over the Internet. Unprecedented amounts of data could now be aggregated and massaged to discover things about individuals not imaginable before. Bizarre, often unnerving, things occurred. Businesses were able to determine that a young woman was pregnant very soon after she discovered herself, and her parents found out through the large numbers of mailings advertising baby items inundating the mail box she shared with her parents. In the future, this will be common place, but now, it is still surprising.

In looking at the issue of the massive amounts of information that are available through business and government, most have not become fully aware of the fact that there are really three different versions of the Internet. The first is the surface web, that which we commonly search through when we do our Google searches. The second is the Deep Web — in which incredibly massive amounts of information are indexed in various websites, usually databases, that are maintained by governmental and business organizations. These are not hidden in the same sense that elements of the third version of the Internet are hidden — in what many would call the Dark Web. This second version of the Internet is just not as commonly indexed as the surface web that we are so commonly comfortable with.

The Deep Web is indexed in various different locales. For example, some of the ways in which you can nibble around the edges of the Deep Web include:

- The WWW Virtual Library
- Stumpedia
- Archive.org
- Science.gov
- Wolfram Alpha.

Science.gov, for example, searches through over 2000 different governmental databases that might otherwise go unknown for the vast majority. To get an idea of the scope of the Deep Web, I strongly recommend that you visit some of these sites and play around a bit. You'll be amazed at the scope and variety of information that is available.[1]

In reflecting about the massive amount of information available in both the Surface Web and the Deep Web, we should, at this point, pause and take note of a couple of major trends that have come into even more intense focus over the last few years. The first of these trends was brought into stark manifestation because of the Covid-19 pandemic. What was a huge amount of business being done on the Internet became a tidal-wave of business being done as massive amounts of business, work, professional activities, education, and a massive list of other activities all moved online as people went into isolation to avoid infection by the Corona Virus. The amounts of data being generated and tracked became unimaginable as almost all of many people's lives went completely online. The amount of tracking via numerous methods, including cookies, became incredibly, indescribably large.

Skeptics about the issue of how much businesses know about us have been known to say that "let them get that data because I have nothing to hide!" This is so very, very wrong.

The first reason why it is wrong relates to the increasing sophistication of machine learning algorithms — algorithms for predicting values, algorithms for discovering data clusters, and algorithms for discovering general patterns in data. The more data that they get, the more accurately they can make predictions about you and your behavior. In many instances, it's information about you that you may not know yourself.

The biggest problem with all this data is that the people about which it is being gathered have very little idea that it is being gathered about them. They have virtually no control over data that describes them. Many businesses regard the data that describes the life of a person as their personal property because businesses gather it and store it. Most people who use the Internet don't realize that there is a gray market of data where some businesses sell what they have gathered about people to other businesses. So, one's own descriptive data, like a child ready to leave the nest, has a life of its own.

The difficult part of this living data is the fact that it can be as personal as you can possibly imagine — including your financial information and your medical data. The implications are scary.

What happens when this organic environment of data and information gets things wrong — as sometimes happens? Some of the outcomes can be very, very bad. For example, what happens if a medical records coder someplace gets it wrong and records that a particular person is HIV positive? Given the living nature of data, once that misinformation becomes part of the data ecology, it becomes very hard to completely expunge it. Medical authorities may strive to keep private information private, but there are so many ways in which data can be uncovered that it is genuinely frightening.

There are so many sources of insecurity: human weakness, technological, and simple sloppiness are a short and non-comprehensive list. Let's consider each.

One could write volumes on the different ways in which people are weak, but it's not always active weakness that's the concern. A lot of the time, it's a genuine desire to do well causing people to do things revealing private information. It can be as simple as a receptionist allowing a flustered "job applicant" to insert a jump drive into a workstation thereby releasing a virus into a network thereby compromising thousands of private records.

Regarding technological weakness, it's probably redundant at this stage to do a comprehensive recounting of common tech vulnerabilities. However, consider this regarding the sensitive nature of medical information. Recent reports describe the fact that computer controlled medical instruments systems are extremely ridden with viruses.

Human sloppiness is another topic that could consume volumes. One hears too many examples of workers being sloppy with someone's confidential information. As an example, there's the story of clerk's leaving personal financial information out where any passers-by would be able to see the sensitive information.

All of these examples serve to emphasize the fundamental fact that data is alive and very, very mobile. Given this fact, what can we do to protect ourselves? Unfortunately, our options are limited, but there are things that one can do to limit the amount of information that leaks into the hands of businesses wanting to know the most intimate details of our lives. What about protecting ourselves from identity theft and other abuses of our personal information? Well, we can do a little more about this.

Interestingly, even though there are two questions, the answers and results of the two are intertwined. If you have better control over what information leaks out about you, the less likely you are to have your identity stolen and the less likely you are to be a victim of a hacking attack.

Most of the answers to controlling the information that leaks out about you are not sexy. Indeed, many of them come from the simple application of common sense.

In controlling the data that we bleed on a daily basis, the first thing that I would remind everyone is the simple fact that there is no such thing as a "free lunch." If you don't pay for it, there is a cost extracted someplace, somewhere, sometime.

The first example that I would draw to your attention is the issue of business discount cards — which are to be presented each time you complete a transaction. The cool thing with these comes from the pennies, dimes, and sometimes dollars that get knocked off your bill when you pay it.

The bad side is the fact that your "ding dong" fetish is now a data element that can be recognized by data clustering machine algorithms at some point in the future. Could this impact your ability to get insurance at some later point? Why? Because your unhealthy dietary habits make you a poor insurance risk. While this direct connection between your grocery history and health insurability, may not happening at the time this paragraph is being written, it is not a question of if but a question of when.

So, realize that a complete data transaction has taken place. The businessman has paid for the right of keeping and using your data and making inferences about you and what kinds of things you are likely to want to buy. Having grown to adulthood before the advent of this data ecology, I, like so many others, don't think about the data and how it can be used against us. Interestingly enough, today's young adults usually don't even seem to think twice about the nature of the transaction and the fact that a little tidbit about them has now gone out into the wider world.

Some might take this concern about a discount card as being alarmist. These individuals would point out that many organizations that collect information about people during transactions also provide a "privacy policy." However, you've got to realize one thing about the limitations of privacy policies: they are a one-sided document that can be revised at any time by the individual or organization that issues them. Fundamentally, there is nothing that prevents someone who has gathered data from you and who has subsequently mined information from that data from selling either the data or the derived information. So, when you give data, consider it having been given to the world!

Getting information from people in exchange for something "free" is a major marketing device on the Internet. Many Internet marketing specialists preach the virtue of giving things away for free in order to establish a relationship with the web surfer. The nature of the relationship being established is to get information about the surfer to help build the mailing list of the marketer. Many different kinds of things are sold in this way: how-to articles, scaled down versions of software, samples of the product or service, and so on.

The key element in this kind of transaction is: how important is this free thing to you? The same admonishments apply to these kinds of services as apply to discount cards. When you give data, it becomes theirs. Privacy policies are not binding to the grantor. People have signed up for things in this manner only to find out that their data has ended up in some rather surprising places. Let me emphasize, you should still read privacy policies — so that you can get a sense of the kind of people who are getting your data and so that you can get a sense of the risk that you are running. Just don't put anything more than minimal faith in these policies being binding.

So, if you are going to protect yourself and your data, the first thing that should guide you is a supreme reluctance to give data about yourself out — for any reason. But, what about the involuntary data collection that comes from the simple act of surfing? We all know these little bits of text that come our way as cookies. Amazing amounts of information can be gathered about an individual based on the cookies that are given to their web browser as a result.

Even after years of dealing with computer systems and their implications, I still find it a bit unnerving to be looking at the result of a search one moment only to find a pop-up ad for that thing or something similar to it suddenly appear on a totally unrelated site that I go to later on. Amazon does this kind of cookie tracking very effectively.

However, social media sites tend to be exceptionally effective in using cookie tracking to gather more and more information about end users. As has been mentioned previously, Facebook's intent is to gather, organize, and package end user data for companies subscribing to their marketing services. Make no mistake; we are not the customers of Facebook. Instead, we are the product.

With our use of Facebook being paid for by information about our likes, our interests, and our personal information, what can we do to limit the amount that Facebook invades our lives and makes us vulnerable to other, even less scrupulous denizens? The first part of the answer comes from carefully watching our account settings on Facebook. Carefully think over what information you post on Facebook and what the accompanying account settings happen to be. I would not recommend posting extensive personal information that can be used to identify you. I'd avoid making my complete birthday available. I'd control who can see my posts. I'd limit the number of apps I use because Facebook has no control over how these third-party applications make use of your data. While Facebook may be careful with your data, these companies have notoriously not been careful. What makes it worse is Facebook is not particularly careful about who it permits to develop applications.

Additionally, since Facebook is all about gathering your data and about providing marketing opportunities, you should be aware that it also is notorious for using cookies to track the online activities of its users. So, this leads us to the general issue of cookies.

Cookies, little bits of text sent from server to browsers, have numerous purposes on the Internet. One of the most common is to keep track of configuration information for sessions when interacting with a website. So, one generally can't use the option that is available with browsers to refuse all cookies, and having your browser ask you permission each time a server wants to send you a cookie, ends up being a tedious pain that often results in users relaxing browser security so much that it often ends up being the same as having no cookie security at all.

So, if a person wants to have some degree of privacy while still having a decent surfing experience, what can be done?

I like to use what I would call cookie management software. There are several different options available. Two that I like working with are Ghostery and DNT+.

Ghostery provides the end user with lots of information. One of the features like about it is the fact that it will identify the various different cookies and widgets you encounter on the web as they are downloaded on your browser. The beauty of this is you can find out who and when somebody is tracking you.

You also have the ability to determine whether or not you'll permit a particular tracking cookie.

I really value the chattiness of Ghostery. Discovering what each web page sends me provides me with an insight to the people whose site I am surfing. I also value the ability to research information about each particular cookie — and the current version of the privacy policies they are operating under.

When I first started using Ghostery, it was not available on all browsers. I initially used it with Firefox. When looking at data from Ghostery, you notice some names that appear frequently. Some of the names include Comscore Beacon, Google Analytics, Double Click, and QuantCast. There are lots of others.

One thing that should be mentioned regarding Ghostery is the Evidon, the company making Ghostery, also sells data collected by its test users to marketing companies. This appears to be a perfect form of synergism. Individual users get their privacy protected. Aggregated data collected by Evidon goes to assist in organizing marketing efforts.[2]

A really good feature of Ghostery is the way that it allows the end user to get rid of LSO or locally shared objects. LSOs are very similar to regular HTTP cookies but were developed to aid in the functioning of multimedia programs such as Flash. The fundamental problem with LSOs is they created some serious security problems in the programs using them. Ghostery explicitly looks for and removes LSOs.

DNT+ or Do Not Track Plus is the product of Abine Corporation. Do Not Track is part of a line of security products that Abine offers. While the other products are pay services, Do Not Track is freeware.[3]

When you install Do Not Track, it, like Ghostery, installs a control button next to where you enter URLs. When you click on the DNT+ control button, it will show you the names of the tracking cookies and widgets on that page. DNT+ blocks all cookies by default. In contrast, Ghostery requires you to choose to block cookies.

One of the really cool things about DNT+ is the way that it interfaces with social media. It will stop a cookie on a social media page and rebuild the originating button on the fly and still allow you use the button.

Besides using cookie management software, other things can be done to manage cookies and your privacy. For example, one thing that you can do is to manually remove cookies at the end of a browsing session. Additionally, you can also configure your browser to automatically remove cookies once a browsing session is over. However, these approaches are not as effective as using cookie management software because cookie users have the ability to track you during a session. So, the decision to dump cookies automatically or manually after a session will not interrupt the session tracking part of cookie use.

So, the wise user of computer systems will realize the nature of the transaction that's being offered by the Internet. Largely, websites are provided by the various entities on the Internet with an expectation of profit. The profit may be the ability to further a point of view. Sometimes, it's the ability to send you ads and help to sell you things. All too often, however, you and your information are the product, and the information gathered is the payment for the cool game or your ability to connect with friends. In those cases, I approach these websites with a great measure of cynicism, and I try to provide as little usable information as I possibly can. I post a minimal amount of personal information on the Internet, and as little as I can of information that couldn't be found in a city directory. In fact, Facebook has less information in its profile of me than does the city directory where I live. Since by posting personal pictures on the Internet I am providing a secondary license to the website to use them, I don't place personal pictures on Facebook.

The issues of privacy policies are just too nebulous. While privacy policies can provide a general sense of the nature of a company, reliance on them is foolish. Companies change them at their convenience.

Most companies gathering information on the Internet simply aggregate data to make more effective marketing decisions, but that doesn't mean there aren't instances where information gathered remains associated with the individual. Remember the incident where the parents found out their daughter was pregnant because of her buying patterns. Simple data aggregation and inference can be done incredibly easily

In the end, it's not Big Brother that's so scary. It's the organic nature of our data and the fact that we have so many watchers, some good and some bad, all watching us. Our attempts at privacy are doomed, but that should not keep us from doing our best to manage what others can do to crawl inside our lives and inside our minds.

While much effort has been expended in this chapter discussing the role of cookies in managing your privacy profile, there are other issues that should be considered in keeping the amount of information you leak onto the Internet to a minimum.

As discussed previously, use of free tools and social media are going to create a direct pipeline by which people, good and bad, will be able to get information about you. So, you will want to pay attention to the privacy controls that you use (who can see your posts, for example) while in programs like Facebook. You'll also want to exercise caution in putting information on social media that could be used to complete a picture of who you are in order to steal your identity.

So, in working with social media, be very cautious about posting any kinds of details about your family or anything else that could be used to complete a picture in order to steal your identity. One way to find out how much you are leaking is to do a vanity search with both your name and your address. Make sure you use quotes to help hone in on results that are directly related to you and your address.

5.2 Special Issues Relating to Facebook

At this point, since we are focusing on the issues relating to social media, we should focus on some issues that are particular to one of the largest, if not the largest, social media tools: Facebook. Facebook is a colossus of the social media world. So much of our national dialog starts and stops with Facebook and Twitter. However, many prefer the more expansive space available for Facebook posts. So, it appears to be considerably larger than Twitter.

When looking at Facebook, one needs to realize that it is not just one company. Facebook owns several different very important social media companies including, but not limited to, Instagram, WhatsApp, and Occulus — the Virtual Reality company, and this was before the transformation to Meta.

So, Facebook, with more than two billion users has an incredibly long reach. Instagram has more than a billion users. WhatsApp has more than two billion users. All of these sites employ cookies and AIs to constantly watch our activities.

Think of the number of applications that you use where you can use Facebook to login to the application. There are more than a few, right? Again, we can see how the reach of Facebook gets extended in so many ways.

Think of the kinds of information that people put onto Facebook. They put their birthdays, phone numbers, addresses, family member names, pictures of themselves, when they are checking in to places and event, and names of friends — just to name a relatively short list of the things that go onto Facebook.

So, Facebook inherently has a massive accumulation of information about pretty much all aspects of our lives — whether we want it to or not. Remember, we might not share a lot of information, but there are no guarantees regarding our friends. Our friends on Facebook can do lots of things that will not only violate their privacy but ours as well. We have no control over the applications that they use in Facebook. Remember, too often the price for use of an application is access to your Facebook data, including your friends list. So, our data travels to locations and is placed in the hands of people we might think twice about or we might actively object to. But our friends don't even consider the implications of their actions in playing that Facebook game, causing our data to be sent even further into the world.

Every time I see a post on Facebook where one of my friends has taken a quiz, for example "what Ralph Roe would look like as an armadillo," I roll my eyes and realize that yet another place has seen my name as an entry from that person's friends list.

Combine this with cookies being used to track us and with AIs constantly watching us, you can see how continued membership on Facebook is anathema to any hope of limiting what Big Tech corporations are learning about you. It is for these reasons that many who are savvy regarding technology and its use are closing out their Facebook accounts and are advocating others do the same.

People have created some interesting terms to talk about Facebook's reach and the broad access they have to people, their friends, and their data. Increasingly you will hear people say that Facebook users have been "Zucked."

The many ways in which your data bleeds away from you: The leakage from your activities on the Internet can take place from many different directions. Some of the directions from which one can have leakage will include blogs and your own web page – should you have one. For example, if you have your own web page and domain, pertinent information about you such as your name, address, and phone number will be available on the Internet about you as part of the domain registration process. There is a neat little web page utility called Whois that is designed to return information about the registrant of a particular website.

The ways in which we leak information are so varied. One very popular tool popular with users of smart phones is Foursquare. Foursquare allows you to use your phone to find out if you have any friends nearby. This is done by using the Foursquare website while on your smart phone to "check-in" to different locations. While there are definite social benefits, this check-in provides additional information to individuals who would track your movements. A nasty outcome of this kind of information has been more than one person having been robbed because of the information posted about their movements on social media such as Foursquare and Facebook.

People tend to have an excessively rosy image of social media. Many will point to the role of social media in aiding socially progressive activities such as Iranian anti-government protestors and such as the rebellion fostered by the Arab Spring.

Evgeny Morozov, writing in his book "The Net Delusion," points out that while the Internet may foster social activism; it also provides a ready resource for the "bad guys" to be able to gather intelligence on activists. The question must be put forth here: why can't identity thieves and other criminals use social media to their advantage against the average person?

In addition to controlling information leakage, it also helps to try to make one's self a little less appealing to potential hackers. We all like the idea of cool Twitter handles and cool Facebook identifiers. The problem with being cool is that cool can also appeal to hackers. While most hackers are going to spend their time paying attention to larger, more juicy organizational targets, it doesn't mean that the individual is immune to potentially very unwanted attention.

Outside of cookies and information leaking, good passwords are essential to protecting your privacy. One of the problems with information leakage is that it often provides a lever for criminals and other desirables to be able to infer your passwords. Too many individuals use the names of pets and family members, often posted on social media, as passwords. Consequently, many experts will recommend that you use something more complicated than a name or a word found in the dictionary. In fact, one of the more common attacks by someone trying to break into a system is to sequentially go through all the words in a dictionary to find out which one will allow access to whatever it is they want. Prosaically, this particular approach to attacking an account is called "the dictionary attack."

5.3 Issues Relating to Passwords

So, one of the major things that many recommend is the creation of more complicated passwords or even pass-phrases that do not appear in a dictionary. Unfortunately, there is a problem with more complicated passwords to protect privacy. People don't use them because the increase in complication almost always makes them harder to remember. You see, some of the standard methods for making things more complicated just don't seem to bind well with human memory. Special characters (&, #, ∧, @, !), when permitted, are often easily forgotten. The use of numbers can help, but again they are often easily forgotten, and if they are associated with publicly available information about you, they really don't add that much to security.

So, the outgrowth of these concerns about passwords is that many security experts recommend the use of pass phrases. With pass phrases, one can keep the individual words simple and the whole easily remembered, but you can also protect yourself from the all-too-common dictionary attack.

To help simplify your life, you may want to think about prioritizing the websites and the security that you apply to them. Try to apply the most complicated and difficult to break passwords or phrases to those websites and applications that have the most sensitive information — banking, shopping, and the email accounts associated with them.

Some people solve their problems with passwords by buying and using a little notebook of webpages and passwords. While this is a better solution than using the same password, your dog's name, on multiple sites, it does suffer from the issue of what happens if you lose it or if some set of unsympathetic eyes comes across your booklet.

If you have lots of different activities online, you may want to consider the use of a password manager. Password managers are sometimes called password vaults. They encrypt your passwords and keep them under control of a program where you can protect them with a strong passphrase. A huge advantage of this type of tool resides in the fact that it allows you to create stronger passwords or phrases without having to incur the burden of memorizing a whole plethora of passwords.

These tools will have a whole range of features. Many of them will provide you with automated tools for filling in forms. Some will permit the ability to randomly generate passwords or passphrases.

Password managers are available for Windows, Mac OS-X, iOS, and Android. Choose one that fits your operating system, has the features you need, and will work with your applications. Password managers are a great way to enhance the security of your work on the Internet.

There are a number of commercial password managers available at a varying range of prices. Manufacturers such as Kaspersky have a commercial password manager for sale. Freeware password managers offer premium versions with advanced security features such as available multi-factor authentication. Multi-factor authentication is where you use something in addition to a password to gain access to a system. Some factors can include biometric measures such as fingerprints and retina scans, or it may include a secondary question only you would know the answer to.

An open source password manager is also available. KeePass, like Linux, is licensed under the General Public License and has versions that will work with

Windows, Mac O/S, Linux, and BSD Unix. There is a pretty large community devoted to the support of KeePass.

In addition to KeePass, you should probably be aware of some alternatives that fall into the realm of freeware. Freeware is a type of software that will allow you to use an individual license for free, but if you use it for multiple people or in a commercial environment, you have to pay for a license.

One program with a free version worth considering is Bitwarden. Besides the free version, Bitwarden has individual and business licenses available and you are not restricted from using it on multiple different kinds of platforms. It's ok, for example, to have it on your cell phone, your laptop, and your desktop — all at the same time. This is not true of all free password managers.

Personally, I am a huge fan of password managers. They save a lot of hassle regarding having to memorize passwords, and they allow you to have big, complex passwords. Many will auto-generate complex passwords for you, and they will often autofill forms for you. Also, by filling passwords for you, they provide a small additional protection from someone trying to log your keystrokes.

So, while the risks of privacy leaking are growing, there are increasing numbers of stronger tools that can be used to help minimize the leakage. We simply need to be aware of their existence, and we need to be smart in how we use that wonderful tool that is the Internet.

5.4 Adventures in the Post-cookie Internet

Although a lot of the discussion has focused so far on the issue of cookies and how they are used to track us, we should be aware that cookies are going to become less important over the next few years. In making this statement, we need to be aware of some issues relating to cookies.

The first issue that we should note is the fact that there are several different classes of cookies. Some of the different categories of cookies include: session cookies, permanent cookies, third party cookies, and zombie cookies. We'll look at these in the following paragraphs.[4]

Session cookies are basically temporary cookies that are used by websites to help web pages recognize a particular person and their preferences as they navigate the various different pages that are part of a website.

In contrast to session cookies, a phenomenon called permanent cookies exists. They are permanently assigned to a particular browser and they often resist regular cookie removal methods. The cookies created by the program Flash are an example of a particularly persistent cookie. Flash cookies are particularly problematical in that they often introduce additional security risks for users of computer systems receiving them.

Zombie cookies are a type of cookie that get recreated even after the cookie has been, in theory, deleted. Flash cookies are often of this type. The reason for these kinds of cookies is that they are often employed by Internet-based games as a way of preventing users from deleting cookies to try to cheat.

Third-party cookies are given out by sites that have the mission of trying to track users across multiple sites around the internet. These are particularly invasive. Third-party cookies are going to be a major focus for discussion in the upcoming paragraphs.

Before going further on cookies, you should be aware that the law regarding cookies is pretty direct. Users going to websites must be informed when a site uses cookies and why they store cookies. Websites must get the user's consent in order to store cookies on a device.

Google has declared war on third-party cookies. They have announced that their browser, Chrome, will soon stop permitting the use of cookies from third parties. By the time that you read this, this change in the way in which third party cookies are handled may very well have been fully implemented.

So, the magic question is this: what will replace these third-party cookies? The answer is something called Federated Learning of Cohorts. This is a fancy way of saying that instead of using cookies, Google will be putting users into similarity groups based on a dynamically created profile developed as you are observed around the Internet.

It should be noted, at this point, that this approach being proposed by Google is not unique or new. It is already used extensively by Facebook. However, despite the fact that it is not really a new concept, many are really upset with this idea. Let's examine why.

Instead of using cookies, Google will be deploying AIs to watch us across the internet. The creation of the Federated Learning of Cohorts is not going to cut down on the invasion of our privacy. Instead, it simply changes the mechanism by which the monitoring takes place. Instead of being tracked by cookies dispatched to our browsers that can be removed, individuals will be watched by Google AIs and placed in a new group each week for purposes of targeting advertising.

Part of Google's plan for implementing Federated Learning of Cohorts involves changes to tools such as the popular Chrome browser created by Google. Google's updated browser will no longer permit third party cookies employed to track users across the internet.

So, what does this mean for the individual user? It means that privacy violations will pretty much continue unabated. It also means that Google will have an even stronger hold on the targeted advertising business. So, it really means the amount of information available about you on the internet won't change, but it will be held in somewhat smaller, though still too large, group of hands.

5.5 The Really Nasty Role of Internet Service Providers

No discussion of tracking and big data would be complete without a mention of the increasingly nasty role of internet service providers in gross violations of your privacy. In a change of policy that took place at the beginning of the Trump Administration, Internet Service providers now have the authority to gather and sell your internet surfing history. Most, if not all, now do so.

This substantially expands the ability of Big Data to track you. Consequently, you really need to consider the use of tools such as VPNs and TOR.

VPN stands for virtual private network and what it does is create an encrypted connection between you and the website you are using through your VPN provider. The creation of this encrypted connection does a couple of things for you. First, it protects the content of information you are sending to the location.

The second thing that it does for you is that it helps to limit the ability of the website you are interacting with to see things like your location. The two things done by VPNs act together to enhance your privacy.

In doing these two things, VPNs limit the damage that your ISP can do to you in attacking your privacy. What the ISP now sees is simply the data flowing between you and your VPN provider. Your surfing data becomes pretty much opaque to them.

So, the question at this point becomes, which VPN? Frankly, I would stay away from free VPNs. They tend to be slow. They may have a limit on the amount of free data that you are able to use before you have to pay. You may also end up having to read through large amount of advertising.

There are a lot of good articles about VPNs available on the web discussing things like prices and performance. I would read extensively and choose something having the features that I want for a price that I feel is affordable. The major players in the market, like Nord VPN and IPVanish, both seem to be solid choices.

Now, there is one little detail that you should be aware of. Your VPN provider can see where and what you surf. One of the dangers of some of the free VPN services is that they have been known to sell your surfing history.

The other major choice for protecting your privacy against your ISP is the TOR network. TOR stands for The Onion Router. It is based on a network of volunteer relay sites scattered around the world. TOR encrypts your data and hides your location from the sites that you surf. It is a powerful tool for enhancing your privacy.

One problem that comes from TOR is that it can slow your surfing down significantly because of the combination of relay points your information is routed through. The big impact of TOR is that you can appear to be from Germany, Sweden, or some other country when you, in fact, are in the United States.

Before leaving this subject, I'd like to offer a tidbit that really is unrelated to security but that you might find amusing. Tools like VPNs are great to allow you to see different kinds of content on various entertainment services. A VPN will allow you to choose the country that you appear to be from. In this way, you have the opportunity to see content based on country of origin you might not otherwise be able to see. I find this kind of cool.

One last one last thing: where you book flights from can sometimes impact what you have to spend on an airfare. Tools like VPNs can really be of assistance in that environment.

5.6 Conclusion Regarding VPNs

The landscape of security for individuals is changing, and some of the changes are new enough for it to be difficult to assess their overall impact. The primary assurance that we have is that maintaining privacy will continue to be increasingly difficult. However, tools like password managers, VPNs, cookie managers, and TOR can really help you slow down (not stop) the information bleed.

5.7 More Issues Regarding Cookies

So far, a lot of effort has gone into discussing the role of cookies in managing your privacy profile, but there are a lot more issues that should be considered in keeping the amount of information you leak onto the Internet to a minimum. Remember, we have multiple objectives at this point:

- Protecting our privacy
- Making ourselves a smaller target for hackers and cybercriminals
- Not giving out information that can be used to attack us.

As discussed previously, use of free tools and social media are going to create a direct pipeline by which people, good and bad, will be able to get information about you. So, you will want to pay attention to the privacy controls that you use (who can see your posts, for example) while in programs like Facebook. You'll also want to exercise caution in putting information on social media that could be used to complete a picture of who you are in order to steal your identity.

So, in working with social media, be very cautious about posting any kinds of details about your family or anything else that could be used to complete a picture in order to steal your identity. One way to find out how much you are leaking is to do a vanity search with both your name and your address. Make sure you use quotes to help hone in on results that are directly related to you and your address. For example, if your name is John Jacob Jingleheimerschmidt, make sure that you search for "John Jacob Jingleheimerschmidt" so that your full name is matched. Otherwise, leaving out the quotes, gives us all results for John, for Jaccob, and for Jingleheimerschmidt. However, I suspect that Jingleheimerschmidt won't add that many hits to your search results other than some references to song lyrics.

5.8 The Problem with Pictures

Cookies are just one issue. Checking-in is another. ISP monitoring is still another. All violate our privacy, and can speak to criminals and hackers. However, there is so much more. Pictures. If cookies speak, then pictures scream, and they reveal a lot more than cookies do. The act of posting pictures is, in so many ways, so very destructive to your privacy and to your security.

The first issue that you need to be aware of is geotagging. The great majority of pictures are now taken with cellphones. Cellphones and many cameras automatically attach metadata to the files that make up your digital images that describe your physical location when the picture. So, when you post pictures on social media, you are telling whoever where you physically go and do things. You may or may not be comfortable with this. However, I wouldn't be.

You see, the problem that we have here is this: who is the audience for your social media posts? Unless you have the security of your social media very carefully set, any images can be seen not only by your friends but by the general public, including lots of unsavories. Remember, it is not just the people who can see your pictures — it is the tools that they can bring to bear on manipulating those image data files.

5.9 Image Harvesters

As if the problems that have just been described regarding pictures aren't sufficient, the act of posting images of yourself and your friends and family is made far worse by the fact that developers have created bots that have the ability to amble through the 'net, through social media and other such creatures, to gather images that are posted on these forums. Not only do they gather images, but they associate names, often tagged in the images by "friends" with particular individuals.

So, when you post pictures of your friends and relatives, you are making it so much easier for those who would invade our privacy and possibly steal from us, and when you tag someone, you are making it ridiculously easy for the snoops to have an image and to know who it belongs to.

Posting images, tagging people, and checking-in make it so easy for privacy-invasive marketers to invade your lives and it makes it easy for criminals, hackers, and other unsavory people to get a picture of you that can be used to figure out how to steal from you.

How about another example? Consider someone trying to steal your identity. Many of the security questions that are used to validate you with different services are based on things such as family, friends, pets, and personal interests. By posting your activities and images, and by tagging people in these images, you are giving potential thieves the information that might allow them to be able to steal you accounts, and to steal your identity.

5.10 The Problem of Embedded Location Information in Digital Pictures

Frankly, I don't post pictures of myself or my family on the web. One of the big reasons behind this comes from the existence of EXIF data inside digital image files. EXIF or Exchangeable Image File format data can be readily used to derive the location that the image was taken.[5]

EXIF file data viewers are all over the web. Tons of tutorials telling you how to figure out where a picture was taken are all over the web. If you value your privacy or those of your family members, don't post your images online.

Yes, you can remove EXIF data relatively easily. If you have to post images, at least figure out how to take the EXIF data out of your pictures before posting them.

The EXIF factor is the reason why I generally counsel people to skip dating services. They are horrible about removing EXIF data, and more than one person has been stalked as a result of using dating services.

At this point, you might very legitimately ask: what about removing EXIF location data? Would that make me safe?

No.

Don't forget that image harvester bots are all over the web. You post images and they are likely to be pretty quickly scooped up by image harvesters. Remember, tools like Facebook and Google already track your location and it is simple for them to associate image data gobbled up by harvesters with location data.

Personally, I have not posted family images on the web, and I truly feel parents who do so, before their children are aware of the issues and can make their own decisions, are genuinely doing their children a disservice.

Recommendation: If you can't live without posting pictures of yourself, your family, and your friends on social media, please consider obtaining an EXIF editor. These are the data elements that are usually embedded in the image files and can include a whole lot of information in them you just don't want to share, including the physical location the picture was taken and the time of day that the image was created. This is not, by any imagination, a comprehensive listing of the information in the EXIF metatagging embedded in graphics files, but it's enough to be worrisome.

Tons of EXIF editor programs exist. They are often available for free to download, and there are competent ones that you can use via the web. Competent EXIF editors are available in both the Apple and Mac stores. EXIFtool, an open source EXIF editor, is available for Linux as well as for Mac and Windows.

5.11 The Process of Hacking

At this point, we should pause and consider the nature of our adversary, the person who wants to steal from us. These people may use a lot of technology or they may use a very little technology. There is one common element in all of this: their desire to take something from us. To simplify differentiating among different types and classes of these individuals, we'll use a shorthand here. We'll call all of them hackers. In this discussion, this term could include regular thieves, people who want to be hackers but who do not have the skills (also known as "script kiddies" or "code monkeys"), and master hackers.

People who work in the security field have identified a process that generally takes place for hackers or other thieves to plan an attack against you (obviously, we are not talking about random street crime at this point). This process will be described here as the five phases of a hack. As presented on a website called "Geeks for Geeks," we have the five phases of hack:[6]

- Reconnaissance
- Scanning
- Gaining access
- Maintaining access
- Clearing tracks.

Admittedly, not every attack will involve all of these steps, but many (or most) will start with the initial phases.

Why this is important and why it is mentioned here, is simply due to the significance, in the attack process, of gaining information about the individuals you expect to attack. Putting information out there on social media about yourself, your family, and your friends is doing part of the work for people who might want to attack you.

While you are not going to stop information leakage as you use the Internet, one of the best ways of protecting yourself is to make yourself as unattractive as possible. An element of making it harder is to not make information about yourself harder to find. Don't volunteer the data. Make the people who would violate your privacy and your potential attackers work for it. Working for it is a huge discouragement. Most simply go away.

5.12 Social Engineering

As has been seen, the use of your information against you is a very common vector for attacking people on and off the Internet. One extremely successful method starts with the gathering of information using soft methods such as data that is self-volunteered. That method of attack is called social engineering, and it has an incredibly successful track record in compromising people and systems.

Social engineering is a wide range of social interaction techniques that are designed to get individuals to provide information or access to information that a requestor would not normally have. Social engineering involves use of basic techniques of social psychology to help judge the receptivity to requests. It involves using techniques of suggestion to help get what you want. It involves working with basic human instincts toward cooperation and helpfulness to be able to get what you want.

Given the breadth of different things that can happen, one can write an entire book devoted to the subject of social engineering. So, what we'll try to do in the upcoming paragraphs is go over a few of the more common issues that might be faced by workers in dealing with the public on a daily basis.

One commonly occurring social engineering attack originates through email. Many variations of occur, but one common theme that happens is when you get an email, signed by the organizations "support team" asking for confirmation of details about the user's account so that service to the account can continue without interruption. The individual receiving this letter, wanting to help the "support team" shares critical account information such as an account login ID and password. In one small innocent act attempt to help, access to an individual's computer, network or database gets provided.

This is but one example of social engineering. Unfortunately, social engineering is a "soft" threat most individuals tend to discount. So, it doesn't get the attention that it deserves.

The scenario that has just been provided is so very common. I can't tell you how many times users have come to me asking questions about a request for information that just came from our "support team." My response, given so very often, is we would never contact you in that way. Naturally, the information requested from the user was a confirmation of their login and password — everything needed to be able to access the network using legitimate credentials. This approach to access is described as "phishing" — a variant of the real word fishing — and it describes the heart of the process very well.

There are so many variations of this attack. One of the more common ones is to send you an email with loads of interesting content and with interesting links embedded in them. Don't fall for this kind of a ploy. It is very common to have these links take someone to a location that runs a program to install a virus or other malware onto your computer. These kinds of email embedded links were part of the tools used by Russian hackers in the notorious 2016 hack of the Democratic National Committee.

5.13 Using our Best Instincts Against us

Corrupting the human instinct to be helpful is a very common approach to getting non-legitimate network access. People want to be helpful. It's a trait that has been built into us over long years of evolution and social development. So, predators will play on these inherent human traits to get them the access they want. Let's look at some common scenarios to see how they commonly play out.

A computer service technician shows up who the receptionist has not seen before and asks for access to the network wiring closet. Knowing that work time for everyone else in the company might be lost if the network is down, the receptionist reluctantly agrees to let the technician, dressed just like the regular technician, into the wiring closet. In the wiring closet, the phony technician attaches a key logger to the keyboard for the main server. A few days later, someone else has all the passwords for the network of the small business.

In addition to the receptionist not wanting to put the other users at the business out of work should the network not be functional, there are other things at play that would have the receptionist help the unknown technician. The receptionist has probably already been told to be as cooperative as possible to the network technician because the manager or owner has already emphasized the high expense of keeping the technician waiting or of having the technician make another trip. So, circumstances cooperate to have the receptionist be helpful.

Like so many of these scenarios, this circumstance can be prevented by being aware of this avenue of attack and by training one's staff to be resistant to this attack. Simply reminding clerical staff that access is never to be given unless you know the technician in question or unless you make a phone call to the company from which the technician is from to verify his identity. That call for verification should only take a few minutes and should not require the technician to make another trip.

The prescribed answer for situations like this resides in a statement made famous by Ronald Reagan: "Trust but verify." Individuals should always verify the claims of people coming your way and asking for your trust. Verify the claims of that hypothetical technician. Train workers to verify. Arrange for your personnel to know who the contact person within the organization is for technical support. Make sure that they know that all requests are to be filtered through that person and that person is the only one to provide information such as passwords and logins to anyone purporting to be a technician. Combine this approach with a call-back validation method. Any requests of this sort are only to be provided after a phone call back to an agreed-on number at the headquarters of the company issuing the service. Controlling access and layering defenses will be common themes that you see repeated in security.

Another variation on how our instincts to help can be turned against us is the scenario that I call "the poor lost soul." Under "the poor lost soul" one has an individual turn up who is desperately in need of help. Maybe she/he has applied for a job at the business next door and their resume has gotten stained. They desperately need the job or so the story goes. Would it be possible to insert their USB data stick into your computer so that a fresh copy of the resume can be printed?

Unfortunately, when the USB data stick is inserted it has a program set to run automatically that loads a virus onto your network. The virus is designed to watch keystrokes and send login and password information to some remote location for nefarious purposes.

The solution to protecting yourself and your organization assets, yet again, resides in training. Under no circumstances should a person from outside the organization be permitted to insert a USB memory stick into one of your computers, especially a computer attached to a network. Good network engineers will often use a tool called Active Directory to disable all USB drives on computers attached to the network. The result of a randomly inserted USB drive can be an infection that moves from machine to machine, compromising the entire organization.

Non-insertion of USB drives very hard to enforce. First, it goes against fundamental human instincts. It's really hard to simply say no to someone in need, but balanced against the needs of the entire organization, one simply has to say no. Now, you can engage in an alternative way of handling things. It would be possible to provide a machine, separate from the company's network, which can be used by these random "poor lost souls." However, that seems to be a very poor use of resources, especially for relatively small businesses.

One crude, but effective, approach for attacking a network involves simply leaving an unattended USB memory device laying around for somebody else to discover. This kind of primitive attack method is often referred to as the "road apple" in honor of something left by a horse on its passage down a road. The attacker is counting on natural human curiosity to get someone to pick up the storage device and to insert it into one of the company's computers in order to find out what it is on the drive. The result can be that a virus gets automatically loaded on the computer and the network.

There are a number of ways in which this type of attack can possibly be deflected. One method involves, wait for it, training. Reminding users that they are not supposed to insert unknown USB drives into computer systems would be the lesson that would need to be imparted. Another method for protecting against a curious worker from attaching an unknown USB device into an organizational computer would be to have technical staff disable USB drives on the organization's computers. This would have the advantage of working consistently but at the price of inconveniencing the users of the organization. This manifestation of the eternal tradeoff in security, usability versus security, would have to be carefully reviewed.

A repeated theme in this discussion has been the issue of training. Training inherently has problems associated with it. The foremost problem with training comes from it being one of the things most commonly set aside by businesses, large and small, when money is short. The benefits of training, particularly security training, are not easily seen. However, these benefits are seen when a network breach occurs, when information is lost, or when a PC or network gets infected with a virus.

Additionally, training, to be beneficial, has to be repeated. Positions, even in small firms, turn over. Policies that exist only in manuals fade in importance. People would rather help than adhere to some abstract set of written instructions—unless they are reminded of their importance.

Beyond wanting to be helpful and being curious, human beings are vessels of weakness and opportunity for potential hackers in other ways. People, even the most hard-headed and pragmatic, are suggestible. Hackers will often play upon this simple fact of human nature to embed things that they want a person to do as seemingly innocuous off-handed comments or questions. Approaching a person for access to a wiring closet or server room, the comment, "I'll bet it would be unusual to let me into the server room," might be offered. Variations of the same theme might be mentioned throughout the conversation with the idea that a seed might germinate and the idea of access might become more attractive. This technique is often referred to as neurolinguistic programming or NLP. It works surprisingly often.

Quid pro quo is yet another approach to getting access to a computer system. Simply stated, an attacker would try to gain access to a computer system by making a member of the organization feel obligated to them, by giving them a reason to help. There are so many ways in which obligations can be created: validating a parking stub, providing information that will help with a fantasy football team, and keeping someone from getting into trouble with the boss are all examples. The critical issue is that a person in which the obligation has been created has to feel the obligation; they need to value what is given them.

Quid pro quo resides in the realm of a gray area between being helpful and being dishonest. It is profoundly difficult to say where one ends and the other begins.

The fundamental problem you have when dealing with human beings is that we don't just leak information. We radiate information. People who want to gather information about us have so many different paths to use. Watch our daily habits. Sift through our garbage (if you don't own a high capacity cross-cut document shredder, get one now).

There are other subtler ways in which human beings transmit information. Our faces and voices exude information for those skilled enough to read it. Much research work has been devoted to the subject of "micro-expressions."

One of the leading experts in the field of micro-expressions is Paul Ekman who helped to formulate essential early research on the concept of micro-expressions. His work has helped to identify micro-expressions as small involuntary facial expressions that can occur when someone is consciously trying to conceal their emotions. These expressions can occur most frequently when there is something at risk in a situation or encounter.[7]

The simple act of observing individuals, businesses, and the ways in which we go about our business. Much of the information that we inadvertently produce is rooted in the fundamentals of human nature. Locking down this aspect of humanity is impossible. My recommendation would be to remind workers to be aware of the presence of unusual individuals in the work environment.

One of the best ways to protect oneself is to try to achieve a degree of security through obscurity. Be boring. Make the act of observation as tedious as possible. Another problem with social media that has not been previously discussed is the fact it can act against one's security by making individuals more noticeable. We all like the idea of cool Twitter handles and cool Facebook identifiers. The problem with this is cool can also appeal to hackers.

So far, in this chapter, we've examined a series of additional threats to one's security and privacy. We've discovered how our friends on social media can make us a bigger target. We've learned how seemingly innocent practices such as posting images of yourself, friends, and family can lead to gross invasions of privacy. We learned about something called EXIF metadata and how it can reveal one's location. We discovered that providing such prolific amounts of information about us can make things so much easier for hackers and for others targeting us because, in the stages of hacking, the first steps are devoted to gathering information. In that same vein, we've discovered how social engineering can use our best instincts against us to help hackers and cybercriminals gather the information they need to attack us. We started to understand some of the ways in which we leak information as we go through the regular activities of the day.

The best we can do to protect ourselves is to slow the leakage and to make it as hard as possible for people who might attack us to gather information needed to formulate an attack. So, as it turns out, the two best self-protectors are awareness of our propensity to leak information and a fundamental caution about adding to the information trail.

But, at this point, we are just pausing as we move into even thornier issues of privacy. One of the big ones revolves around the issue of constant location tracking that we are subjected to.

5.14 The Thorny Issue of Location Tracking

First, let's start with the bad news. This one is really super hard to beat — if it can be beaten, which I doubt. You have way too many factors going against you. However, you can make it harder and you can lessen its impact.

5.14.1 The Problem of Cellphones

A major part of the difficulty facing each of use resides in the devices that the vast majority of us carry with us. Who is going to give up the convenience of cellphones? In those devices, we have the root of what essentially amounts to guaranteed defeat in the constant face of location tracking.

The difficulty that we have is embedded in the nature of the technology that we are working with. Cellphones continually send and receive signals that attempt to locate the existence of a nearby, contactable cellphone tower. People in the technology industries refer to this process as pinging. This comes from two sources: those old war movies involving submarines hiding from enemy destroyers by remaining silent where the sending of a locational signal via sonar was called pinging and the term also comes from computer networking where an attempt to see if a computer exists at the other end of the communications connection by sending out an "are you there? signal" That is also called a ping. The term PING is an acronym that stands for Packet INternet Groper.

So, inherently, most of us trying to hide our locations are going to be doomed to failure in that the cellphone pinging process is going to inevitably be connected to programs that will send our locations back to the mothership of the various companies that make the software that drives the operation of our smartphones. At this point, there are two main companies that make the software for smartphones. They are Google and Apple.

So, Google and Apple, pretty much always know where you are.

This is why many who are very concerned about privacy continue to hope for the implementation of Linux-based cell phones. At the point of writing, some have just appeared on the market, but their efficacy has not really been established.

There is also another issue to be discussed: de-Google-ization. We'll address that issue a little later.

5.15 The Basic Problem of Traitorware

Several years ago, an organization called the Electronic Frontier Foundation identified a new type of software that it described as Traitorware. The basic definition of Traitorware can be summarized as follows: software in devices that acts against the interests of the user. Things such as location tracking and the location information embedded in digital photographs are examples of Traitorware functionality.

The genesis of Traitorware as described by the Electronic Frontier Foundation came from patents applied for by Apple to allow them to protect hardware and software by allowing them to track and disable devices remotely.

Apple's basic idea was pro-customer: trying to track down and/or disable devices that had been stolen.[8]

However, this tracking of stolen devices was not the only application of these technologies. These technologies can also be used to "brick" appliances that users are trying to "jailbreak." At this point, we need to define some terms: bricking and jailbreaking.

Bricking is just that: permanently turning the appliance into a non-functional inert collection of electronic circuits.

Jailbreaking is a relatively important issue for iPhones. However, a version of it can be done for Android phones also.

Jailbreaking refers to the process of removing manufacturer restrictions on the software that you can install on your device. Some of you may find these restrictions problematical. The basic logic is this: I bought the device, I should be able to install anything that I want on it.

You may, however, want to think about Apple's side of this issue. Putting restrictions on what you can install so you can only install approved applications from the Apple Store is actually generally going to be to your benefit. Keeping the applications on people's iPhones that are exclusively approved by Apple generally limits the chance of malware being introduced onto your device. This is part of Apple's "closed garden" approach to ensure a high level of customer experience.

There is another issue that you may want to consider: jailbreaking voids your terms of service that you agreed to when you started using the device. Some would say: who cares? Well, you should. The reasons why come from the issue of software updates. If you jailbreak your phone, you will break your terms of service agreement, and you probably will eliminate your ability to get software updates.

From my standpoint, this is a huge issue. Despite Apple's claims to the contrary, iPhones do get viruses. Updates are super-important to help protect you against vulnerabilities. Completely updated or patched systems are the first line of protection against computer viruses. Not receiving updates basically severely compromises your security. The ability to run some little passing app is way less important than the position that you put yourself in by not getting updates.

Another issue you may want to consider is this: jailbreaking is not technically illegal at the time of this writing, but it has been illegal in the past. This existence of jailbreaking resides in a grey area of legality and is not something that I care to traverse, and it is largely due to the rules of something called the Digital Millennium Copyright Act. The DMCA is a colossally bad piece of legislation, but that is beyond the scope of this document. However, one interpretation of the DMCA is that jailbreaking and related activities harm the rights of the manufacturer and are consequently illegal.

5.16 "Jailbreaking" Android

The proper term to use here is "rooting." This refers to the idea a user getting superuser or root administrative access to the device to be able to install pretty much anything that you want.

Why would a person want to root an Android device? Often, specific models of smartphones are associated with special offers from different cellphone providers. Getting access to new, cool tech and then moving to a less expensive service is appealing to some. Frankly, I wouldn't do it. The advantages of a supposedly hot phone are relatively transitory, but the consequences of violating terms of service and not getting updates will exist through the entire life of the device.

Frankly, my approach to this problem is to purchase unlocked phones and to make sure that the phone is encrypted, has anti-virus on it, and uses a VPN.

5.17 Privacy and Location Issues with both Android and Apple

To restate: both Apple and Google track the heck out of you. Although I am not an Apple user, I have more sympathy for their reasons than for those from Google. Google's reason seems to primarily stem from their overall objective of being able to analyze the heck out of you in order to sell you more stuff. Apple is very highly probably also guilty of that, but they do seem to have the additional motivation of being able to track and potentially permanently turn off, or brick, the device.

Outside of the issue of tracking you, Apple does seem to be relatively committed to protecting the privacy of their customers.

5.18 Location Tracking Apps

Many applications track you. Cell phones have applications built in to them to make your location known. I would generally recommend that you turn off these location tracking tools. However, it appears that keeping the location tracker offline is hard to do. Many applications need locational data, and it appears that some of them cause the location tracker to be turned back on. It seems that keeping location tracking inactive is relatively hard to do. You will need to check your settings periodically if you do not want your location shared.

5.19 Just Don't Check-in

As discussed here previously, checking-in online is a bad thing. Social media encourages people to do this, but more than a few people have been the victim of crimes as a result of posting their check-in, only to come home to find that their home had been burgled. This is one of the most direct manifestations of the danger of location tracking.

5.20 Limiting the Impact of Manufacturer Tracking

Location tracking is really hard to beat. Both Google and Apple do it extensively. There is really little that you can do to stop them. However, there is something that you can try. It is called de-Google-ization. Because Android is derived from Linux and is open source, Google's influence and information gathering in Android can be limited to a degree. If you are strongly concerned about Google tracking you, it is possible to acquire what are called "de-Googled" phones. Several sellers provide de-Googled phones.

Other styles of smart phones exist. One alternative that is coming on to the market at this writing are Linux-based cellphones. Linux is the penultimate open source program. Hiding surreptitious information gathering is going to be very difficult to do. I'm looking forward to the broader availability of Linux-based cellphones.

5.21 De-Googling your Life

De-Googling your phone is just one action that you can take to try to limit the degree to which Google invades your privacy. There are a series of other actions that you can try that will slow Google down in its tracking of you. Notice how this was phrased: "slow Google down." Again, we need to recognize that, short of selling everything, cutting up our credit cards, and moving to live in a drafty cabin in the Montana wilderness, there really isn't a lot that you can do to genuinely preserve your privacy in the world as it stands and as it is likely to develop.

Location and other forms of tracking engaged in by Google have so many different dimensions that it is really impossible for the individual to defeat. However, here are some suggestions of things you can do to help you reduce your information exposure:

- Get and use a de-Googled phone. They are available and will cut down on the amount that Google is able to track you.
- Use a VPN. A big part of the location tracking comes from regular Internet surfing. VPNs will allow you to choose where you appear to be from.
- Alternatively, use TOR. TOR will route you through a random combination of volunteer servers. You can appear to be from Holland one day and from Sweden another.
- Periodically, if feasible, randomly turn off your cell phone. They can't ping your phone when your phone isn't on. Turning off your cell phone inserts more noise into the tracking data, thereby making the data less reliable for the trackers.
- Use search engines that do not track you. This includes Qwant, Swisscows, and Wolfram Alpha.

5.22 Conclusion

Location tracking is pervasive. The fact you carry a cell phone is going to mean you can be tracked. However, you can limit its impact by the choices you make. Turn off location tracking on your cell phone and deny it when applications on your phone request it. Don't check-in on your social media and applications. Recognize that location tracking is not completely bad. Apple uses it, at least in part, to help protect your privacy if your phone is stolen.

Chapter Notes

1. "How are 'The Dark Web' and 'The Deep Web different?" Lyndon Marshall, Practical Insecurity, accessed 24 November 2022, https://practicalinsecurity.com/?p=131
2. "Ghostery: A Web tracking blocker that actually helps the ad industry," Venture Beat, accessed 6 December 2022, http://venturebeat.com/2012/07/31/ghostery-a-web-tracking-blocker-that-actually-helps-the-ad-industry/
3. "Blur (formerly Do Not Track Me) for Chrome," Download.com, accessed 6 December 2022, https://download.cnet.com/Blur-formerly-DoNotTrackMe-for-Chrome/3000-2144_4-75653397.html
4. "Different Types of Internet Cookies," Millie Johnson, Rocket Lawyer, accessed 6 December 2022, https://www.rocketlawyer.com/gb/en/quick-guides/different-types-of-internet-cookies
5. "How a Photo's Hidden 'EXIF' Data Exposes Your Personal Information," Thomas Germain, Consumer Reports, accessed 6 December 2022, https://www.consumerreports.org/privacy/what-can-you-tell-from-photo-exif-data/
6. "Five Phases of Hacking," Geeks for Geeks, accessed 6 December 2022, https://www.geeksforgeeks.org/5-phases-hacking/
7. "Microexpression," Wikipedia, accessed 6 December 2022, https://en.wikipedia.org/wiki/Microexpression#
8. "What is Traitorware?" Eva Galperin, Electronic Frontier Foundation, accessed 6 December 2022, https://www.eff.org/deeplinks/2010/12/what-traitorware

CHAPTER

6

Training: Your Best Protection

This chapter is going to focus on issues that are important to business. However, we will look at some issues important to individuals at the end of the chapter.

One of the ongoing themes in previous chapters has been the issue of the importance of training. The effects of so many problems can be limited by a training yourselves and your employees. There are several levels of training that should be considered. These levels include basic security awareness training, basic network training, and basic training on how your systems are configured. Let's consider all three after we consider some possible sources of training.

Some different sources of training for you and your organization include in-house experts, consultants, online training, and experts from local colleges and universities. Each will have their own strengths and weaknesses.

In-house experts, when an organization is large enough to afford them, are certainly cost effective and accessible ways of getting training. However, some negatives exist that managers should also focus on.

In-house experts tend not to be highly regarded. There is that inevitable contempt of familiarity one may have to contend with and that may limit the impact of the message one is trying to convey. External experts often inherently have more validity in the eyes of many.

Another factor to consider is the fact that shaking in-house staff loose for the event can be problematic. IT staff tend to be overworked. Most organizations view IT as a cost center, and they tend to try to run their operations as cheaply as possible. Putting the additional burden of training on technical staff can result in a none-too-memorable event when a harried IT pro puts pressing operational concerns over an educational session.

That observation about IT manager priorities said, education of users should be an integral aspect of the responsibilities of an IT leader. Many IT leaders make it a point to read extensively about the newest security threats, and will often share that information with computer and network users in their organization through email-based newsletters. Some news providers provide newsletters for recirculation among end users.

An IT leader making it a point to communicate with end users can help to ameliorate potential problems within an organization. Reminding users about common attacks such as phishing can help remind users to not click on that link or to not reply to that email with login information.

In addition to proactive communications, the effective individual supervising computer or network operations should also be quickly reactive to threats as they appear. When an end-user reports a phishing attack, a technical leader should be very rapidly responsive to the situation and should communicate through emails, phone messages, texts or whatever format the organization is comfortable with as soon as is possible. Don't let users drift. Speed in communication can help to forestall the activation of a possible network vulnerability.

Consultants are another source of possible training for training. Since consultants in any form represent an additional cost, most organizations will only consider the use of consultants when a specific pressing immediate need has been identified. Ongoing training is probably not a good candidate for consultant-led efforts.

Identifying consultants to lead training will be problematic. Good technical skills and good communications skills can intersect, but they often don't. Any consultant you consider should provide some evidence of both technical and communications skills. Some indicators of technical skills will include degrees, diplomas, and certifications. Additionally, a curriculum vitae should show a history in which these skills have been applied. Don't forget, there are individuals in the field of technology who might be appropriately described as "paper tigers." They have degrees, diplomas, and certifications, but they have never had to practice their trade in earnest.

As always when making a significant investment, you should try to investigate the success that others have had with the individual. A list of previous customers is in order.

There's a fundamental element of the consultant's equation that managers often neglect and consultants would like you ignore. Consultants have a standardized, pre-packaged set of services and perspectives they offer you. Don't expect any real customization of their services. Either their standard services will fit or they won't. It will be up to you to determine whether or not those services come close to your needs. While they will do their best to convince you that you need what they have, you need to question whether or not they have delivered or can deliver the kind of educational services needed by you and your company.

With online training, your alternatives grow substantially. Some of the most exciting alternatives for training reside in the realm of online professional education for technical staff. As of the writing of this document, some of the most exciting venues for education in technical areas reside with Youtube, Coursera (and other non-traditional educational start-up initiatives), MITx, Udemy, Teachable, and other prerecorded educational programming.

You can find nearly anything that you want on Youtube. It varies in quality from excellent samples of highly professional educational training programs with first class production values to kids showing you how they fought a "boss" on a particular game.

Many commercial training companies will upload samples of their courses on Youtube. Given the fact that commercial training can be so expensive, this option often provides a good opportunity for someone to try a particular training "on for size."

So, a question that we might ask at this time is how does one evaluate the quality of a particular Youtube provider? First, you will have to sample their work. Do they seem to have a mastery of the subject matter that they are presenting? What are their production values? What kinds of comments are being offered related to the video? Are other presenters saying similar things?

Consider the nature of the videos being offered in the channel. Do they present issues that seem to have importance when you do searches in that field? Are there a wide range of courses being presented in the subject area? Are they focused?

New forms of education and training are being developed all the time. Several educational companies now advertise themselves as alternatives to traditional education and training. One of these initiatives is Coursera. Coursera provides a wide variety of courses in a variety of disciplines. This includes courses in biology, engineering, psychology, and computer science, including security. Their courses include pre-recorded, interactive lectures, homework, assignments, and discussion boards for individuals having problems.

MIT has had a long history of providing free educational resources. Their Open Courseware initiative has provided engineering and scientific study materials for many years. They have started with interactive courses similar to those of Coursera. Their interactive offerings are now housed under an organization call EdX, which includes, at writing, 12 different universities such as Harvard, University of California at Berkley, and the University of Australia. Courses in a wide variety of subject areas such as the social sciences, technology, and science have been made available to the public for free enrollment. For the most part, the courses are not for credit.

You may want to consider courses from sources such as Teachable.com and Udemy.com. Both provide short focused courses in various areas of interest, including many related to general IT subjects, including security. Teachable courses are a little more expensive than those on Udemy, but both have offerings that are pretty well focused on specific areas that might be useful to the individual or to the businessman interested in learning how to better protect themselves or their business. Looking at these resources, following the tradition of a lot of online commerce, user reviews and reactions can give an idea of the quality and the nature of the focus of the course.

Some of these options, such as edX, MIT and Coursera will be more useful for IT professionals. One of the key elements which will continue to focus the importance of security training are the problems encountered by everyday users. One of the key exposures that organizations continue to suffer is the silly things done by the untrained end-user. Coursera and EdX focus more on high-end technical education while Teachable and Udemy will focus on short courses in subjects that user training might come from.

One thing to consider for training that is more targeted to everyday users would be to see what help local colleges and universities can provide. Many will have continuing education departments that may be of help to small businesses and individuals with technical literacy issues.

The help you can get from colleges and universities may take several different forms. Some will have evening community outreach programs. Others may have special community outreach events where tech literacy issues can be addressed. The best way to pursue this option is to contact the school's outreach or continuing education officer.

I am an increasing fan of Udemy. Udemy has an extremely broad range of courses in a huge array of fields. You can find advanced learning courses for just about any technical field you can imagine. However, more important to this discussion is the fact that you can find a large number of courses that are designed for the beginner, including areas such as networking and Internet basics. Just doing a quick search of the site, I found some interesting introductory materials in basic network security that would be useful for training the individual needing to know what's going on with a computer network. There were also some interesting learning packages introducing things such as how the Internet of Things works, for example.

What is really nice about things like Udemy is prices are generally attractive. You can get nice introductions to things such as Office or the basics of the Internet for something that is generally less than what you would pay for a book on the subject. So, this option might be workable for the individual wanting to learn. Moreover, the company also has attractive corporate training packages that could be employed even by small businesses as an element of their fundamental training repertoire.

6.1 Basic Network Training

Many under estimate the importance of basic network training to the safety and security of individuals and small businesses. A large proportion of the problems that organizations have with computing come from stupid things that people do with computers. Basic tech training can help ameliorate that stupidity. The worst thing that can come from educating users as to computer and network operations is fewer support requests. This can ease the personal frustration of the end user, the burden on clerical personnel with IT responsibilities, the amount of support that a harried IT department has to provide, and the number of visits that have to be paid for outsourced IT support.

Sadly, organizations go cheap on basic network training for their personnel. Why? Because they think that they have it covered when they put requirements in job ads that list competence with Word, Access, and Excel. Many businessmen don't realize that basic computer literacy does not mean network literacy.

So, what can be done? A couple of approaches are necessary. First, your local computer people need to be involved because every network setup is different. Your workers need to understand your network and the shared resources that it provides. You need to make it a part of company ritual to provide new workers with the introduction they need to how your systems operate.

Regarding a second approach, one should note that the formal organizational introduction may not be enough or it may not be relevant for the single computer user or solitary business person. Given those circumstances, you will often need to seek out additional network literacy training.

Where do we go for basic network training? Without a discriminating eye, online videos such as those available from Youtube would not seem to work very well. The preponderance of videos I've found in poking around would be good for technologists or aspiring technologists, but not good for someone just wanting to know the basics so that they are a little less dangerous.

Looking at websites for sundry book sellers, self-education titles seem to be relatively available. So, this might be a workable alternative for the individual or very small business.

While you may also consider consultants, one should refer to the previous discussion of the implications and limitations of this source of training. Usually more effective, probably more available, and certainly less expensive would be the short courses that many colleges and universities provide on the basics of computer and network literacy.

6.2 The Individual

But what about the individual person wanting to understand a little more and protect themselves a bit better? As mentioned previously, Youtube might not be the best option for someone in this situation. Although, a search at the time of writing found a few possibilities, Youtube is not where I would turn to deal with self-education in this situation.

Doing a generalized search for information on networking and internet basics, some fairly decent options appear. One is from an organization called GCFLearnfree.org. This organization provides a pretty decent introduction to networks and the internet at the following URL: https://edu.gcfglobal.org/en/internetbasics/what-is-the-internet/1/.[1] Additionally, they have a pretty large library of instruction on using things like MS-Office. They also have an internet safety course oriented toward children.

To conclude, training will not be a cure-all, but it will certainly be a cure-lots. So many security problems and general network problems start with the user.

Chapter Notes

1. "Internet Basics: What is the Internet?" GFC Global, accessed 7 December 2022, https://edu.gc fglobal.org/en/internetbasics/what-is-the-internet/1/

7

Email — The Ongoing Flaw in Our Armor

In my experience, all of us have a strangely bifurcated relationship with email. We both love it and we hate it. In addition to grumbling about the overall unbelievable stupidity of management, we continue to grouse about the amounts of email we get. Having been the designated local administrator of a locally hosted exchange server, I got to see the overriding weirdness of our relationship with email. One thing I found interesting was the reactions that people have to spam (not the canned meat-like product). For example, we complain about the junk email that fills our in-boxes, but we continue to engage in behaviors that are guaranteed to cause an absolute avalanche of unsolicited email offers. Again, as yet another example we continue to sign up for offers where the only price for admission is our email address, and we are not particularly discriminatory of where we scatter the seeds of our email address. Yet, when the seeds we have tossed hither, thither, and yon produce the bloated fruit of the continuing expansion of garbage in our in-box, it seems to continually surprise us. Email eats our days — so much so that we have productivity gurus telling us how to escape the iron embrace of email in our lives. The uncounted hours we spend in rushing to respond to spurious emails has often been counted one of the banes of modern life.

This realization of the corrosive nature of email to the functioning of our personal and professional lives is recognized in numerous different ways. Some organizations will declare "email emancipation" days where workers are expected to completely empty their in-box. Books on productivity are written that preach putting response to emails at the end of the day or that preach strictly limiting the time that one spends on email.

If only these productivity issues were the only problems with email. Beyond these time-wasters, there are so many ways in which email can cause security problems they are hard to enumerate. Some of the different problems that can come your way from email include: phishing attacks, viruses hidden in attachments, and viruses embedded in links in the emails themselves.

7.1 Getting Hooked Through Phishing

We've discussed phishing before, particularly in the chapter focusing on, among other things, social engineering. To restate, phishing occurs when someone directly asks you to send them some sensitive information that can be used to hack your network account, empty your bank account, or steal your identity.

The classic example of phishing seems to be a cottage industry in some parts. In one common variant, one is asked to send money, a bank account routing number, or something similar to a displaced prince or relative of a prince who needs your assistance to regain access to millions of dollars that have been tied up for various different reasons. Of course, you will be given a substantial reward if you help the poor benighted royal. Variants of this request became so common across the Internet that they became a meme, a cause for general laughter. Ironically, people continue to fall for these sad attempts to swindle money or information from them.

As noted previously, a common form of attack against organizations and individuals is to pretend to be from a technical support organization. This may be in the form of a social engineer pretending to be a technician, but it's more likely that the form of the attack will be an email where the sender pretends to be from "tech support" or from the "support team." The usual gist of the message is to ask for the user's password. I can't tell you how many times I have gotten emails from users of my networks asking me if these requests were genuine — after I've repeatedly told them that I would never send them an email asking for their password in that way. More than one person never seemed to attach themselves to the fact that, as network Administrator, I could simply log-in, change their password, and then log-in as them.

Note that in making these requests for information, the attacker often already has some knowledge of the end user: their user name. This adds a degree of credibility to the note. Remembering that people want to be helpful, that they need to be helpful, one can easily see how these kinds of attempts to get users to compromise their information would be successful more often than one would like. In this process of using their username to establish credibility, many users naively forget that tools like Google Dorks (not kidding—that is a real name) can be used to scour places inside the surface web or the deep web to reveal things like user names and email addresses.

Next, when these would-be information pirates add in a sense of urgency, it makes it harder for the target to resist being helpful. The sense of urgency can take a number of different forms. One is to inform the user that their account has been compromised and he will lose access to it if one doesn't change password by following the link in the provided email. Of course, in responding to this urgent appeal, the user has now placed the password in the hands of the attacker. Many users can't resist this form of attack. Succumbing to this form of attack isn't a measure of stupidity. I've seen many very skeptical appliers of extreme critical thinking skills succumb to them.

Again, as mentioned previously, awareness training can help to cut down on the success of this class of attack, but these attacks really do prey upon the best instincts of people. So, complete protection would probably require a complete revision of human nature. Having worked as an IT leader, I can't count the number of times that I've had to reassure users that these mails were, in fact, false and that our department would not contact them in that way.

What makes phishing attacks really dangerous is when they are combined with other forms of intelligence gathering to make them seem more plausible. How could someone get the name of the company's owner, or the IT manager, or the head of technical support at the company you get your computer services from if they weren't legitimate? The answer is that a very little light research will find all kinds of things that are not particularly well hidden. It's really hard to resist giving information to Joe, you know, Joe — the guy we get tech support from.

This particularly insidious form of attack is called spear-phishing because of the way in which it mimics the act of carefully targeting a fish in a barrel before releasing the projectile. Many of the attacks by the Chinese against western commercial and governmental interests are in the form of spear phishing. Using their own intelligence services, they do the footwork necessary for the attack on technical systems more likely to succeed. They establish credibility by finding out things like user names, names of friends and relatives, and whatever else is necessary to seem real.

Let me emphasize: one does not need the resources of the Chinese intelligence services to successfully launch a spear-phish. Too many sources of information are available about us. Google is now the first choice, but let's not discount the effectiveness of city directories, still available at many city libraries. The hordes of information available on the Internet are accessible in ways many don't yet fully comprehend. The fact is this: specialists in areas such as open source intelligence who are experts at dredging up obscure information, such as background on individuals, are becoming increasingly common.

7.2 Google: Sweet Puppy or Savage Wolf?

We need to pause here to consider the dark side of everyone's favorite search engine. As an integral element of Google's drive to create the world's largest and most formidable advertising company by putting more and more of human information at our fingertips, Google has provided a growingly powerful tool that can be used against you in endeavoring to focus phishing attempts more and more specifically against you, to make emails being sent you by criminals seem even more credible.

For example, at the organizational level, Google provides tools that can reveal many of the email addresses that are used within organizations. These tools can be used to strengthen the credibility of phishing emails that come your way.

7.2.1 Google Dorks: Another level of exposure...

The concept of the Google Dork is simple: use advanced search methods that are built into the Google search engine to allow you to reveal things that might not normally be revealed in a standard basic Google search. Google Dorks are incredibly powerful. An entire database of different classes of Google Dorks is available for potential hackers or the curious and is easily available over the internet.

The consequence of the existence of Google Dorks is that information that we might normally regard as hidden on the internet is now available and discoverable via the Google search engine. Again, the focus of the phishing attempts become even more direct and specific to their proposed targets.[1]

7.2.2 Our exposure is not just due to the tech...

We don't have to resort to the twisting of powerful tools to increase our vulnerabilities or the vulnerabilities of our organization. Simple observation can often provide very telling evidence. Who provides technical services to the small company? What does a person's daily routine consist of? Simple external reconnaissance often provides that answer. Let's not forget physical mail and bills as a source of information. Everyone should own a cross-cut shredder.

7.2.3 The simple act of inference...

More than one person has commented on the fact that they can tell when there is a military crisis in the United States by sitting out and watching the pizza delivery vehicles come and go from the Pentagon. In a more mundane realm, more than one company has been burglarized because observers saw an empty parking lot, and that fact allowed the burglar to infer workers were away from the business. With the very human inclination to help, email will always continue to leave us vulnerable. For reasons like this, the simple act of putting your car into the garage can aid your security in unexpected ways. Observers cannot infer your absence by not being able to see your car in the driveway. People have a harder time being able to predict your movements.

7.3 Attachments, Detachments, and Vulnerability...

Returning to the realm of issues directly related to email, attachments represent one of the biggest ongoing elements of vulnerability that organizations continue to have. Attachments can mess a person up in so many different ways that it is genuinely hard to enumerate.

One big way in which you can be hurt as a home user and as a business is through macro-viruses that can be hidden inside word processing files, spreadsheet files or database files PDFs, and so on. In fact, any common application program can camouflage a virus waiting to attack your computer or your data files.

To refresh memory, a macro-virus consists of a script of keystrokes and commands meant to automate common features of the program currently being used in order to do harmful things to data or to the computer. Commonly, email clients will warn you about the possibility of embedded viruses in attachments. Take those warnings seriously. A macro-virus may have been embedded in the file being detached. Remember, there are so many variations of viruses out there that your anti-virus may not have recognized the presence of the virus hidden inside your attached file.

So, never open attachments from someone you don't know. It's an all-too-common form of attack. However, you also need to be cautious about attachments sent to you by people you know. It's all too easy for a potential attacker to pretend to be one of your friends. How can they discover who your friends might be? That's easy. Sites like Facebook are extremely porous. Depending on your security settings, even strangers might be able to see people who you commonly correspond with, and that's just one way in which they can discover potential plausible venues for attempting to trick you into opening an attachment that you shouldn't.

So, what's the defense? Some parts of the defense have already been mentioned. Only open attachments from people you know and from which you are expecting an attachment. Even better, don't use email attachments at all. Instead, make use of various different files sharing services instead like Dropbox, iDrive, or Onedrive. These companies generally provide pretty good security on anything passing in and out of their computers.

If you find that you have to make use of attachments, make it a point to look at the email address of the person sending you the attachment. Is this your friend's email address? Most phony email attachment attacks fail this simple test.

However, you really should go another step. Contact your friend or associate and find out whether or not they actually sent you the document. Why do this? The difficulty is email addresses can be spoofed. It may look like your friend's email address, but bad actors can make it look like their correspondence came from somebody else.

7.3.1 Death by executables...

There are nastier creatures in the cybernetic forest than those hiding in word processing or spread sheet attachments. While those can hold malicious scripts, they are not as bad as full-blown programs. Sometimes folks will send you zip files or actual executable programs. Zip files are one of more common types of compressed data files that are used for transmitting relatively large files over networks. Compressed data files are to be regarded with extreme suspiciousness.

As noted, Zip files are a type of compressed file. Others include RAR, TAR, and JAR. There are a large number of alternative formats that are used for archiving and data compression. The whole issue in the use of these different file formats is to save space. Many of these file formats were created early in the life of the Internet where communications were often implemented via dial-up. At the commonly available dial-up speed of 56 kilobytes per second, downloads were often painful waiting affairs. Compression file formats were developed to help individuals to aid in data transfers. If you could develop a "condensed" file that took out a third or half of the space of the original, that would be so much better!

The major and scariest problem with file compression, as noted, is that these file formats are often used to hide scripts or programs that, when they are "reinflated" or unzipped, can attack your computer system.

Some of these compression formats are known as self-extracting executables. Rather than requiring you to purchase, download, or install a compression program needed for reinflating the archived file, the data and decompression program are packaged together in something called a self-extracting executable.

This is really a very convenient tool for transporting and decompression, but this convenience brings with it a second edge. The executable may just as well hold a virus inside it.

So, attachments are a convenience that can hold lots dangers for you and your computer. Be very careful in opening them.

7.3.2 To protect yourself, simplify!

E-mail itself is dangerous, and I'm not talking about the bizarre crank chain letters that ask you to bombard your next 3000 closest friends. While that may be scary by itself, most security specialists are more concerned about the possibility of your computer getting infected with a virus.

Most email comes from in a web-activated format implemented by HTML. Hyper Text Markup Language is the primary device that's used to construct the web pages we view on the Internet. These web pages can include pictures, animations, movies, and programs.

All of the elements of the previous list can have viruses embedded in them. Pictures are a frequent vector. Animations are too. Programs, however, are the most dangerous pathway for viruses to come into your computer system.

Many security experts recommend that one of the first things that you should do to enhance your email security is to set your email format to text only. This process will vary significantly among email clients, but information on how to do this is pretty easily available on the web.

Have you noticed how often that emails with embedded hyperlinks will arrive in an email client with the client declaring that it is protecting your privacy and that if you want to see the images and content, you will need to "click here?" This is an attempt, by your client, to implement that protective lack of functionality that comes from using simpler plain text emails.

So, end users are faced with yet another trade off. They have to choose between the enhanced functionality of plain text versus the fact that every email coming into their system may carry a destructive payload with it.

7.3.3 Exposing yourself through email!

It has also been mentioned previously that one should be very careful about signing up for "free" offers. Almost every free offer that you will encounter will require you to input your email address in order to get access to the free service. As has been mentioned elsewhere, this is a common marketing device that Internet marketers use to build their direct mailing email address lists. What's noteworthy here is the fact that they will also usually provide you with a "terms-of-service" agreement that many will regard with a warm and fuzzy feeling that it codifies and legalizes their use of the free service or software.

Unfortunately, most don't read into the bowels of the agreement. If you are like me (and I suspect most of the rest of us), you only skim it partially and impatiently before clicking on the "I Agree" button so that you can get your next new toy. What is actually hidden in the dark recesses of the agreement is your acceptance of the fact that they will be sharing your personal information, your email address, with their "business partners."

By sharing your email in this fashion, you are deliberately lowering your defenses against spam and other invasions of privacy. Be very careful in providing your email address because, in putting your email out there, you are inviting attacks.

7.3.4 Protect yourself using throwaway email addresses!

So, what if you really want the freebie that apparently requires you to give up your personal email address? How do you avoid clogging -up your email with the junk that comes from sharing your email address? In asking these rhetorical transitional questions, remember, when you give your email away, you are giving it not just to this person or site, you are also giving to all the "business partners" of that organization.

The answer is simple: get yourself a throw-away email address to sign-up for free offers. Seriously, there are a bunch of companies that specialize in allowing individuals to create temporary addresses so that their primary address will have less of a chance of being junked-up.

As of this writing, some of the services that provide this help include: 10 Minute Mail, MailDrop, EmailOnDeck, Guerilla Mail, and ThowAwayMail. This list is not exhaustive. However, I do strongly recommend checking them out.[2,3]

7.4 Can Spam? Not!

If you've followed the computer industry even casually, you may have heard of the CAN SPAM Act. However, this law was enacted several years ago, and the continuing rise in unwanted email would tend to belie its effectiveness. The acronym stands for Controlling the Assault of Non-Solicited Pornography And Marketing Act. The act was an attempt to stem the tide of pornographic spam that was and is assaulting the sensibilities of email users. Even the relatively innocuous emails for sexual enhancement drugs are problematic. Having administered an email system, I can tell you that one of the banes of an email admin is the constant inundation of ads for Cialis and Viagra, especially from foreign pharmacies. One of the things that came to amaze me was the ingenious ways in which spammers would find to spell variations of Cialis in "leet-speak" (substitutions of numbers for letters, for example).

One element that is almost always an element of spam is an unsubscribe link. The CAN SPAM act requires that Internet marketers provide a way for recipients of unsolicited emails to be able to opt out of email lists. The fundamental problem with this requirement is the fact that a large fraction of genuine spammers are located outside the reach of American state and federal laws.

Fighting spam, I found myself frequently looking at the hidden information coming with an email that dictates its source of origin. Many of these spammers come from places like Eastern Europe: Romania, what was once Yugoslavia, the remnants of the old Soviet Union, and other places in Europe, like Italy.

So, what's the point? Many of these countries have a Wild West attitude toward technology. The only acknowledgement of US laws is a wink, sneer, or on an exceptionally expressive day, a shrug. They are not bound by our laws.

To make this point even more strongly, let me emphasize. Organized crime is often involved in the distribution of spam. The tools they use to distribute their tacky wares is often directly destructive to Internet users. One common vehicle used by gangsters for distributing their wares is the Bot Net.

Bot Nets happen when a virus infects a computer system and does things that the owner/user of the computer is not aware of. One of these things is to use your computer to distribute spam. Computers, working together in a Bot Net, can do lots of nasty things. One is the distribution of spam. Another is distributed denial of service attacks where computers work together to overwhelm the capacity of a network to remain on the Internet. In net lore, computers that have been compromised to distribute spam or to do other nefarious activities that are unknown to the user/owner are called "zombies."

Being aware of these kinds of issues, I've concluded that it is foolish to try to clink on the unsubscribe links of any email, no matter who it comes from. There are several reasons for this conclusion. First and foremost, clicking on the link confirms for the unethical sender that they have a real live address – one that they can send more nasty emails to. Also, they can take this little tidbit and resell it to other spammers. At one time legitimate email addresses could net you a half a cent on the black market. Realize that many spam emails are generated through random address creation programs. Programs will use different combinations of domain names and user names to try to shotgun their sales pitch to whoever or whatever is on the other end.

One might legitimately ask the question: "Is this paranoia justified when the sender is an American company?" The answer is decidedly a "yes." For example, how can you be sure that the sender is who they claim to be? Often, junk email senders will pretend to be someone else, a friend or a company you do business with. Remember, your surfing habits are constantly being watched. Cookies and other tracing devices are often employed to observe your activities, what sites you go to, who you email, and so on. Many of these observing eyes are decidedly not friendly.

I'm not convinced that the CAN SPAM act or any other well-intended attempt at regulating spam has done any genuine good. Some users have gotten a false sense of security from these laws while believing that the Internet is, and remains, an Americans-only electronic playground. Many continue to maintain this American-centric viewpoint of reality while tsk-tsking over the latest news of Chinese government hacking activities, or Russian government hacking activities, or....

So, block SPAMmers aggressively, and never ever click on an unsubscribe link.

7.5 A summary, One of Many...

E-mail is a huge security gap for every organization. Its protocols and communications methods and pathways are well known by those who would violate your security. It is always going to be a path for possible organizational and personal security compromise. The only way to stop it would be to eliminate it entirely. That's something that's not going to happen any time soon. So, virus ridden, open to spoofing and phishing, it will continue to pose a risk for all of us.

Chapter Notes

1. "Smart Searching with Google Dorking," Exposingtheinvisible.org, accessed 7 December 2022, https://exposingtheinvisible.org/guides/google-dorking/
2. "Five Best Disposable Email Accounts," Aseem Kishore, HelpDeskGeek.com, accessed 7 December 2022, https://helpdeskgeek.com/free-tools-review/5-best-free-disposable-email-accounts/
3. "How to Create Disposable Email Addresses," Jon Martindale, Digitaltrends.com, accessed 7 December 2022, https://www.digitaltrends.com/computing/best-sites-for-creating-a-disposable-email-address/

8

Social Media and Social Reputation

When I first wrote this chapter, I opined, "social media has infiltrated our lives." Looking back, that statement was a weak understatement of what the reality has become, and it is an even weaker avatar of what promises to be.

In the intervening years, we have seen social media become a tactical and strategic nexus of two elections. We have both liberal and conservative elements of our society excoriating technology companies over technology's corrosive influence on society and on our elections. Despite all of this, many of us don't reflect about the pervasive impact of tech and social media on our lives. Individuals who would normally say that they are private people, Tweet their activities in a blow by blow recount of every step of their day. These same individuals upload pictures of their latest meal. They feverishly make new professional and social contacts via social media.

In some ways, it seems as if normal social reticence disappears when it comes to an electronic environment such as Facebook, Pinterest, Instagram, Snapchat, Youtube, and so on. People post pictures of themselves at parties — often without a second thought. When one factors in the images involving alcohol and other intoxicants, the information that people reveal on the internet is amazing

Employers have begun to twig to this chatty propensity on our part. It's now routine for employers to check the social media pages of potential applicants as a way of ensuring that individuals of good character are selected for the job. More than a few reports have been made that some employers demand the social media logins of potential employees as a prerequisite for being considered for a position.

Because of the waxing interest of employers, it is now standard for human resource specialists to counsel job hunters to make sure that they clean up their social media entries before job searching. However, there is one problem with this mode of fixing your online history: things that are placed on the Internet never go away. When considering the crawlers specializing in harvesting images and data, the stuff we place on the web has the can haunt us for a long time. Factoring in websites dedicated to preserving images of old websites, the possibilities for the persistence of web pages long after their destruction is truly remarkable. One well-known site for preserving old websites is the Wayback Machine, now known as the Internet Archive.[1]

One can begin to see the depth of concern about things persisting on the web in the belief that you shouldn't post something that you would be uncomfortable in "grandma" seeing. One of my lawyer friends posted a meme that says, essentially, "dance like no one is watching you, but write your emails and social media posts like your lawyer is watching."

8.1 Freedom of Speech and Slacktivisim

Images are not the only thing that one should be on guard for when posting information on social media websites. Increasingly people attach themselves to and participate in causes through the Internet. Businesses, politicians, and causes commonly market themselves through the Internet. We can see this manifested through the continual marketing messages that implore us to "like" them on Facebook. If you "like" or follow a cause, it is not uncommon to be flooded with requests to sign one petition or another.

Liking causes and signing petitions are sometimes referred to as "Slacktivism," especially when it connects to politics or to a cause. The reason "Slacktivism" is applied stems from the fact that instead of working to support a goal, one's interaction is limited to clicking on links on the web. However, lacking in real commitment it may be, Slacktivism provides real benefits to the recipients of the clicks. Here, you are providing the advertising organization with a free addition to their contact lists. Instead of having to use broad advertising efforts, the click provides a focus point of individuals who have already expressed an interest in the item or service being offered by the organization.

Politicians, causes, and businesses all make very effective use of social media — so much so that a whole new category of professional has evolved in the last few years: the social media marketing expert. Social media experts have been popping up all over the place. In many ways, they are proliferating like computer and web "experts" proliferated in the 1990s.

Providing insights on the quality of social media marketing experts is beyond the scope of this book. However, social media by itself is not going to win a fortune. It needs to be part of a broad range of marketing activities. Successful organizations integrate social media activity into systems that involve various levels of customer support and are frequently labelled as CRM or customer relationship management.

In any event, organizations are very sensitive to what is said about them in various forms of social media whether it is Yelp or Facebook. Many will hire full-time professionals whose job it is to monitor social media for bad things being said about the organization. Some consulting firms provide this kind of service for quite handsome fees.

This sensitivity to social media influences extends to quasi-social media environments such as large online merchandisers such as Amazon. Negative reviews can really impact sales of products. Aware of the potential for harm to sales, there has been more than one instance of companies "salting" favorable reviews of their product at places like Amazon. Of course, plenty of approbation gets heaped on companies engaging in these activities, and this marketing tool has even been discussed at some length in the business media, including blogs in the Wall Street Journal and Forbes Magazine.[2] The issue of fake reviews has been going on for a long time, and it remains a problem for sites like Amazon. The problem of fighting fake reviews was discussed in an article in Time Magazine called, "Inside the War on Fake Consumer Reviews."[3]

While this production of fake reviews might seem to be an abuse of freedom of speech, it certainly isn't illegal or punishable, and marketing specialists, when caught, usually are not punished. Some companies seem to regard being caught as part of the game.

In looking at the issue of false reviews, it should be noted that not all the reviews being salted are good reviews posted directly or indirectly by a company's marketing arm. In some instances, the false reviews being provided are negative ones designed to harm the competition, any competition.

To provide a sense of the scope of the problem, there have been instances of good and bad reviews having been purchased in bulk and on the cheap from online service purchasing websites such as Fiverr. Additionally, when you add in the fact that other individuals have been found to create multiple false accounts in order to use them to create buzz for their product or to harm the competition, this area of online reputation can become quite problematical for organizations. So, two questions arise. How do companies fight bad reviews and how do customers detect false reviews?

When you do a search on buying false reviews, purchased ads extolling the services of companies dedicated to fighting false reviews often pop up. Recently, ads claiming software solutions for detecting false reviews have started to appear. However, as we'll see later, the issue of false reviews is still being studied. In all of the ongoing struggle regarding fake reviews, what is fascinating to see is the number of business articles warning managers and executives against the practice. Yet, it continues.

In terms of detecting the false reviews, there are things that be done to protect oneself. One action that can be taken is to look to see if the reviewer has done other reviews. If they haven't, it can be a hint that this is a purchased review. Again, checking their history, one can look to see if there is a huge predominance of reviews for one company's products or services. This can be a strong clue that they reviewer is employed by the company, particularly if the reviewer has posted negative reviews for the competition.

The web is a big, wild, and open place. There are lots of opportunities for the unscrupulous to try to manipulate the medium. Some will lie or get others to lie of their behalf. Practice skepticism.

8.2 The Virtue of Vanity

Individuals and businesses can both benefit by periodically doing vanity searches. Simply search using your own name. Even without the issue of finding what appears about you, the results are often fascinating. While one can find instances of people saying bad things about an individual or an organization, one can also find instances of folks saying good things or of somebody with a very similar name who one might not want to be associated with.

However, don't dwell too much on what you find via a vanity search — one way or the other. The simple fact is that what others think of us is often beyond our control. That said, it is still better to be aware of a pattern of commentary than not.

8.3 Social Media, Social Manipulation

One of the few things that the very conservative and the very liberal tend to agree on is the fact that they hate tech companies. Conservatives complain that tech billionaires are often "woke" and they (and their companies) tend to be biased toward liberal causes. The very liberal complain about how much leeway people like Donald Trump get on platforms such as Twitter and Facebook. After finally being banned from Twitter for spreading disinformation related to COVID, a change of ownership saw Elon Musk revoke Trump's ban. At the time of writing, Trump has yet to exercise his option to abandon his own platform and return to Twitter.

Some contrarians tend to conclude that they must be doing something right to have both sides angry at you. Alternatively, one could suspect that they are doing something so very wrong if it is recognized as such by people on both ends of politics and perception.

However, it really should be noted that some social media platforms have been caught red-handed doing some very nasty direct manipulation of participants.

In an article in Forbes magazine that was originally published in 2014, the author describes how Facebook participated in being used as a human behavioral laboratory.[4] In this laboratory-itization of Facebook, the platform manipulated the feeds of several hundred thousand subscribers to pump either all positive or all negative information into the newsfeed. What they found was fascinating. An old saw stated that places like Facebook constituted a place where people went and felt bad because they were jealous of other people's lives. What they found instead was that negativity, for example, was contagious. A negatively influenced subscriber would post fewer positive posts, and this trend toward negativity would be picked up by others. Several other outlets, like the Atlantic, addressed the great manipulative experiment as well. One fundamental conclusion was this: emotions are contagious — especially those on social media.

This situation would have been bad enough had it ended there. This little experiment set the stage for a much larger one with probably greater consequences: The Great Cambridge Analytica Debacle.

For those who are unaware, Facebook was the platform that was used by Cambridge Analytica to launch an app that purported to do a detailed psychological profile of the user. Instead of so many of the other games that you see being offered you on Facebook, like Farmville or Candy Crush Saga, this one was directly sold to users as profiling tool.

And boy howdy, did it profile! Pretty detailed psychological profiles were formulated on millions of Americans participating in the application. Like lots of applications on Facebook, the application got the contact list of participant's friends. So, following the information gathered in the 2012 experiment, they had a direct pipeline to impact attitudes of the users of the application and the friends of the users of the application.

To make the problem even larger, Cambridge Analytica was given extended access by Facebook to millions of profiles of Facebook users. An article published by the Guardian in 2018 cites the profiles given to Cambridge Analytica as exceeding 50 million.[5]

The big issue stemming from Cambridge Analytica is that the data produced by the Facebook application and the data given by Facebook to Cambridge Analytica was used to very sharply target political messaging in the 2016 election, and the campaign that benefitted from the actions of Cambridge Analytica was that of Donald Trump. Supporters of the Republican candidate used Facebook and the profile information in the hands of Cambridge Analytica to determine both targets and content of information they put out over Facebook during the campaign.

The issue of Cambridge Analytica is actually a part of a broader issue that pertains to Facebook in particular and to Social Media in general. The issue is simply this: Facebook has historically done a poor job in vetting the applications available as add-ons. They have been very cavalier about the way in which these applications make use of our personal information, and given the fact that these applications almost always demand access to your personal information (name, phone and friends list), this wild-west attitude toward apps becomes even more problematical.

So, when you combine the factors of Facebook directly manipulating people and their emotions, of Facebook participating the manipulation of millions of users by providing Cambridge Analytica with a portal to this manipulation, of Facebook not vetting applications in general, and of Facebook literally handing over large amounts of data to Cambridge Analytica you have plenty of criteria for seeing Facebook as corrosive to people and their privacy.

So, having seen too many requests to play Candy Crush Saga, to play Farmville, or to take a quiz to determine what superhero I look like, I wince in the realization that here goes my information to yet another unreliable source. Yes, I can help solve the problem by dropping social media, but, unfortunately, the proverbial cat has already escaped the proverbial bag.

8.4 I Hate "likes"

Social media and online stores share a fascination with participatory user feedback. In social media, these are the ubiquitous likes that seem to be the common thread that helps to determine what we see in our news feeds. In online stores, the common thread that manifests itself are the number of stars and the attached customer reviews for the various products being offered. Both of these issues create potential problems for us and one could argue that these issues are security related.

In the area of social media, the likes that we bestow on posts help to determine what is pushed in our direction. In using social media, we always need to keep in mind the fact that they are really advertising companies that devote their software to tracking our interests so that they can more effectively sell things to us.

We are the product, and our likes are what is sold to the various advertisers that poke their heads up pretty much all the time through our use of the platform. The direct consequence of this is this constant modification of the digital portraits constructed of us by social media. Another direct consequence of these digital portraits is a kind of "silo-ing" of people into a digital echo chamber where the only things that we see are things that reflect our interests and world-view.

Since many get their news from social media, this process of creating information silos is very problematic. A 2021 research study by the Pew Foundation, found that a significant proportion of Americans got news from social media, that for half the users of Twitter it was their primary news source. Recent years have seen a large growth in the popularity of media pushing particular viewpoints or supporting special interests.[6] Social media appears to have supported this popularity increase. Social media, contrary to rosy myths, does not seem to bring people together, but it seems, rather, to add to social strains and fragmentation.

So, what can a person do about the corrosive impact of user/customer feedback and participation? Well, even though they are superficially similar phenomena, the actions that one would take to deal with the problems created by social media are a little different from what one would do to handle the consequences of fake customer reviews. Since a method for detecting salted reviews has been presented, let's focus on escaping our self-created phenomenon of being held prisoner by what you have "liked." However, before doing that, I would mention that a fair amount of formal research is taking place in computer science regarding the detection of fake reviews, and some good progress has been made in developing software for the detection of fake reviews. However, given the continued large presence of fake reviews, I am reluctant to declare that a solution has been found.

Regarding escaping the strait-jacket of our likes, I don't really have any good news for you regarding potential technological solutions. You are going to have to be actively engaged in seeking out good news sources, and I believe that they do exist. However, one has to be aware that news is often dependent on advertising revenue, and consequently is part of what many would describe as the attention economy. What that means is that what drives the news is often going to be that which outlets believe is most likely to gain the attention of customers, and this is going to be even more true of highly biased news sources that prey upon your outrage — whatever your political preferences happen to be.

Recommendation: Even though a complete book could be written on the issue of media bias and methods for dealing with it, I would offer the following method for loosening the straight-jackets we have created for ourselves:

- Don't get your news from social media. It'll often reflect the perspectives of your friends and those you follow.
- Don't buy the nonsense that your friends and associates circulate on social media, especially if it seems oriented toward making someone angry. Research it. See what non-biased sources are saying.
- Use search engines that don't track you to do your research. Several things will impact what you see in your searches, but two of the biggest are what is currently trending on search engines such as Google and what you have searched for in the past on search engines that profile your search history. A couple of non-tracking search engines that I recommend at the time of this writing are Qwant and Swisscows.
- Familiarize yourself with the Adfontes Media Bias chart. It is a respected source for rating bias of different media sources. Knowing the slant of your sources helps in assessing the quality of the news.[7]

8.5 The Great Balkanization of Social Media?

So, at the time of writing, we have an interesting phenomenon: large numbers of both liberals and conservatives leaving platforms like Facebook for something else. Some liberals, citing the bad history of Facebook regarding privacy, have not really settled on alternative, and are often making the departure from Facebook a significant issue. Conservatives, citing the treatment of politicians on social media platforms depart to alternatives such as Parler (defunct at the time of writing) and MeWe.

First, regarding the division of social media, it seems as kind of an extension of echo-chamber/filter-bubble phenomenon we have been witnessing across our society, fueled by social media in recent years. There are two direct sources to this process of social silo-ization. The first are algorithms. Advertising and our friend suggestions on social media are driven by the assessment of programs of our interests and who we might like. The second is ourselves. We seek out people who we know, who we like, and who are like us.

Secondly, looking at these attempts to create alternatives to the mainstream social media services, I would tend to be pretty conservative about rushing in to join them. For example, when Parler was a popular alternative, and before it closed, it suffered a massive hack of its participant data. By some estimates, 70 terabytes of user data got stolen and ultimately dumped for the interested on the Internet.

The issue here is this: are the alternatives competently programmed? Given the cavalier attitude by most social media organizations to the idea of privacy in general, one really doesn't want their information spread even further through sheer incompetence.

Apparently, the Parler hack was fueled by a very basic flaw of having all data posted on the network pretty much available to just about anybody.[8]

At the time of writing of this chapter, Elon Musk has been going through the spasms of adjusting to the purchase of Twitter. Large numbers of employees in areas such as security and content moderation have been let go, and many observers have been taking bets as to how fast it is going to take for Twitter to completely implode. Many individuals, unhappy with the tenor of Twitter under the Musk regime have been migrating to alternatives such as Mastodon. However, Mastodon has been severely criticized for being difficult to use. The net effect is that, many are continuing to dither regarding their migration to a new platform. It'll be interesting to reread these words in a couple of years.

So, given these situations and the current state of things, I'll be going slow in any potential move to an alternative service.

Chapter Notes

1. "Wayback Machine," Archive.org, accessed 7 December 2022, http://archive.org/web/
2. "Fake Reviews: Amazon's Rotten Core," Suw Charman-Anderson, Forbes Magazine, accessed 9 December 2022, https://www.forbes.com/sites/suwcharmananderson/2012/08/28/fake-reviews-amazons-rotten-core/?sh=75dbcd1c7428
3. "Inside the War on Fake Consumer Reviews," Megan McCluskey, Time Magazine, accessed 9 December 2022, https://time.com/6192933/fake-reviews-regulation/
4. "Facebook Manipulated 689,003 Users' Emotions for Science," Kashmir Hill, Forbes, accessed 7 December 2022, https://www.forbes.com/sites/kashmirhill/2014/06/28/facebook-manipulated-689003-users-emotions-for-science/?sh=496ebb67197c
5. "Revealed: 50 Million Facebook Profiles harvested for Cambridge Analytica in major data breach," Carole Cadwalladr and Emma Graham-Harrison, Guardian, accessed 7 December 2022, https://www.theguardian.com/news/2018/mar/17/cambridge-analytica-facebook-influence-us-election
6. "News Consumption Across Social Media in 2021," Mason Walker and Katerina Eva Matsa, Pew Research Center, accessed 9 December 2021, https://www.pewresearch.org/journalism/2021/09/20/news-consumption-across-social-media-in-2021/
7. "Interactive Media Bias Chart," Ad Fontes Media, accessed 9 December 2022, https://adfontesmedia.com/interactive-media-bias-chart/
8. "An Absurdly Basic Bug Let Anyone Grab All of Parler's Data," Wired Magazine, accessed 15 December 2022, https://www.wired.com/story/parler-hack-data-public-posts-images-video/

9

Wifi, Home Networks, Home Security, and the Internet of Things

---

Everybody has a network at home. Increasingly, people not only use their home networks to surf the web, but they also use it to support remote work, and they use it to monitor the home appliances, children, and pets. They use it to run the preponderance of their entertainment. Increasingly, home networks are becoming integral to our lives. As such, there are huge opportunities for security to be attacked, and it means that individuals need to take more care to try to secure themselves and to try to maintain some shred of privacy. In this chapter, we'll take a look at the particular security problems that are associated with Wifi and home networks, the problems you may want to consider when dealing with home security, and the particular security problems associated with the Internet of Things.

9.1 The Problems with Wifi

It's a drama that is played out in so many homes. Dad changes the Wifi password and informs Junior that he can't get access to the Wifi password until he cleans up his room. It goes to demonstrate how integral these elements of tech have become to our social structures and mores. That being said, you should know that many (I would say most, but I can't back that up) network engineers, don't really care for Wifi and would be just as happy if it disappeared from their lives.

You see, Wifi is a security nightmare. It inherently creates more opportunities for people to be able to attack you. Additionally, it is all too easy for Wifi signals to be interfered with by electro-magnetic radiation created by the grinding away of other electronic devices. So, with the potential for security and reliability issues, Wifi tends to add to the headaches of network engineers.

So, let's talk some basics of Wifi security. First, you need to put a password on your Wifi. Period. No arguments. Not having a password on your Wifi means that others can see and attach themselves to your network. People wouldn't do that! They would do it in a heartbeat, and some poor slobs have been seriously messed over by having some predator attaching his/her self to their Wifi and doing some nasty things like downloading somebody else's copyrighted movie. Guess who got charged with the copyright violation by angry Hollywood lawyers? You got it, the poor slob who did not secure their Wifi.

Look, this practice of trying to attach to somebody's Wifi connection is so common that it has a snarky name attached to it. It is called "wardriving." This is where someone gets into a car with their laptop set to detect wireless signals and literally driving around looking for unsecured wireless signals. Back in the day when we used phone lines to carry Internet signals, we would call the process of methodically calling phone numbers looking for a computer to hook to as "wardialing."

At this point, we need to stop and indulge in a bit of technical terminology. What's really germane at this point is a term called "SSID." SSID stands for Service Set Identifier. SSID is basically the name of the wireless network. They are usually broadcast to help potential network users know when a network is available for use.

So, the stronger a network signal is, the farther that this SSID announcement will be made. On the other side of the equation, a stronger signal means a more reliable connection to the wireless network.

At this point, I should probably mention that SSIDs do not need to be broadcast. Not broadcasting the ID will add a small increment of security to the network by hiding the network from discovery from the casual wardriver. It will still allow legitimate users of the network to connect.

When you decide to put a password on a network, you are deciding to encrypt the connection to the network. This also means mathematically scrambling the information being sent and received so that a person can't intercept and read your information. It also means requiring a password for access.

In encrypting the network, you will generally have several options. The one that you should choose, and at the time of writing it should be available on the wireless routers available for purchase, is something called WPA3, which stands for Wifi Protected Access 3. As implied, there have been three generations of

WPA, and WPA3, by far, presents the highest level of security. If that is an option for you on your wireless router or wireless access point, choose it.

Recommendation: WPA3, just don't use anything else. You may have the option to choose a couple of other wireless networking protocols. Among those available are WEP (Wired Equivalent Privacy), TKIP (Temporal Key Integrity Protocol), WPA, and WPA2.

WEP sucked. It was so insecure that TKIP was rushed for development to try to plug the major security gap created by WEP. WEP had serious operational flaws that allowed hackers to force the revealing of the access key. An even minimally capable hacker could compromise a WEP wireless environment in minutes.

WPA and WPA2 were serious improvements over WEP. However, as often happens with protocols, they don't necessarily age well. As time passes, flaws in the implementation of the security protocol get revealed. This is the case with WPA and WPA2. WPA2 is really long in the tooth. It has been around since the early 2000s. It is really time for a change — so much so that if you have a wireless router that's been around for a few years, it really is worth your while to invest in a unit that supports WPA3. WPA3 is that much better.

Recommendation: Stay away from Bluetooth if you can. Bluetooth is a wireless protocol that is used to hook up wireless devices that are relatively close to what they are supposed to connect to. There are two types of Bluetooth connectivity. Class 1 devices can be 100 meters (328 feet) from the thing they connect to. Class 2 Bluetooth devices can connect up to 10 meters.

Bluetooth has some real security issues associated with it. As a protocol, it is really easily hacked. So, use of Bluetooth increases security holes in your network. Class 2, with its distinct distance limitation, is a lot more justifiable. I find it hard to justify Class 1 devices.

Consider some situations. You might use Bluetooth to connect a wireless speaker to your cell phone so that you could have better sound when listening to your Youtube music videos. That makes sense. One could also use a Bluetooth printer to allow the placement of a printer physically far from your computer. That might very well be a Class 1 device. That would make me squeamish. Pumping out data from your computer to be printed on a printer could allow someone up to 300 feet away to intercept your data.

In contrast, I would have fewer concerns using a Bluetooth speaker to project sounds from my cell phone, for two reasons. First, I would probably be playing music videos from Youtube. Second, using my cell phone with a Class 2 device, such as a speaker is likely to be, is probably not going to expose any more information that is already exposed through my using an Android phone or through my use of the unblinking eye of any service associated with Google.

Oh, by the way, turn Bluetooth off on your cell phone when you aren't using it.

9.2 Home Networks and Digital Assistants

When I think of home networks, I tend to think of PCs. That is archaic thinking. Now, when discussing home networks, one has to include things such as tablets, TVs, and digital assistants – including Alexa, Siri, and the Google Personal Assistant. Furthermore, this issue of digital assistants will also bleed into a discussion of the Internet of Things, but there are some special issues regarding the Internet of Things that demands its own separate discussion. To begin this exploration, let's talk home networks and personal assistants.

Your internet router is your point of demarcation between your home or business network and the wild and wooly world of the Internet. Networking types will often use the term "demarc" to describe the point of interface between one network and another. In talking about connecting one network and another, most business networks will have a separate physical device dedicated to running firewall protection software working along with the business router serving as the demarcation for the business network. However, in most personal environments, firewall software will be run on the individual devices connecting up to the network. My point is this, besides having anti-virus on your computer, make sure that there is firewall protection somewhere in your computing environment. Don't expect protection from your router. All it does is make decisions as to whether or not something stays in the local network or goes out to the Internet.

There are several areas in which home networks provide security and privacy vulnerabilities. One big one is with smart TVs. There has been more than one instance of hackers attacking smart TVs, and in some very notorious situations, there have been instances where the manufacturers of the TVs have used built-in sensors to monitor the goings-on of the rooms in which they are deposited. You know, it's a variation of the old joke, "In the USA, the TV watches you...."[1]

However, despite the problems associated with smart TVs, we seem pretty well dedicated to the fun and convenience of having a menu-driven environment in which pretty much all the content of the Internet is available for us to be able to stream, at the punch of a button.

So, the magic question then becomes this: if these things are going to be pretty much a basic feature of our home networks, what can we do to minimize the security risks that attend having smart TVs as part of our lives? Well, most people who are aware of the gnarly world of smart TVs will recommend the following sets of actions:

- Get dedicated external devices. Devices such as Roku, the Amazon Fire Stick, Chromecast, Apple TV, and others are generally better programmed and are usually more secure than those that are completely integrated into a TV. This is recommended in more than one blog published by security firms, and it also reflects what I have seen in working with smart TVs.

- Always-on features are dangerous. Review them, and turn them off if they are not essential to your use of the TV. Many recommend that a better solution is to turn off and disconnect the set when the TV is not in use. One of the sad tales of tech is the fact that there are almost always undiscovered vulnerabilities inherent in software. Smart TVs, still a really young tech, seem even more prone to these problems. Eliminate an attack pathway by turning off and unplugging your device.

- Turn off the microphone and voice command features of the device. For various reasons (some to be discussed later), these capabilities really make me nervous. Again, more than one instance of vendors using these devices to listen in to their customers has been documented. You don't want the even less benign presence of some hacker listening in when you are in for the evening with your family or significant other.

- Turn off the camera if your device comes with one. There really should be an indicator light, and if there isn't, why have the silly device? The same issues that apply to the microphone apply here.

- Make sure that you visit the part of the menu in the smart TV device that pertains to software updates. Set your device to receive updates automatically. As noted elsewhere, most hacks and invasions of privacy for these devices and for other computational devices come from vulnerabilities that can be fixed by doing software upgrades.

- Review other settings. Some settings will reveal more information than you might be comfortable revealing. For example, the device may send information regarding what you view when it phones home. Despite a possible convenience, I would not be comfortable with that.

- Read your reviews. When in the market for a smart TV, it really is worthwhile to read the reviews and focus in on the issue of what is said regarding the security of a particular device. Start with security as one of the leading requirements when buying your smart TV.[2]

Recommendation: Unpleasant though it may be, I am going to suggest that if you are concerned about security and privacy, you probably want to stay away from the use of personal assistants. This simply stems from the basic nature of what these devices do, and it also stems from the architecture by which they do those things. Let's take each of those issues in turn.

Look, tools like Alexa and Siri control many of the devices that are connected to your home network, and they may control many elements that are accessible in your home network through the Internet of Things (again, more on IoT later). As they do things like turn on your lights based on your vocal command, raise the temperature of a room, turn on romantic music, or start your coffee brewing, they potentially handle a lot of the homey activities of our lives. As directed by devices that are elements of a network, the accomplishment of activities from smart devices also represent data points that can be used to create an intimate picture of the internals of your daily life.

At this point, when I start to rant on this issue, many visit the "I have nothing to hide" fallacy on me. Most of us are not cognizant of the things we reveal about ourselves through the agglomeration of our daily activities. My usual response to the "I have nothing to hide" assertion is this: how do you know that? My take on the whole thing is my daily routine is my business. This verbal control of some very basic elements of our everyday lives is what I mean when I refer to the fundamental nature of what devices like Alexa, Siri, and the Google Assistant do.

The intrusive nature of the data collection by personal assistants has been documented by more than one writer. Things such as your name, your time zone, your phone number, your payment information, your age, your interests, the location of your device, your IP address, a history of your locations are among some of the pieces of information that these devices collect and send back to the mothership. Furthermore, most of these devices will rifle through you contacts and store information on them such as their address, their email address, their phone number, and they relationship to you. This isn't true of all personal assistants, but it is true of more than one.[3]

Still even further, it has been documented that more than one of the providers of these services has recorded things being said in the locale of the device. It is very common for me to have people comment about the weirdness of talking about something and then having that thing feature prominently in the web adverts being funneled at them.

So, if the basic nature of what is being done by smart assistants is problematical, what could make it worse? I would suggest to you that the architecture by which they accomplish their magic makes them even more problematical. Let's remember the basic nature of what these devices do. They perform vocal recognition: capturing sounds and interpreting those sounds as commands in what computer scientists would describe as natural or human languages. Those human language commands are then reinterpreted as commands that the machine (Alexa, Siri, Cortana, or the Google Assistant) can issue to the network connected smart device.

For me, watching the evolution of these devices for an extended period of time, that is an impressive list of accomplishments — simply reinterpreting human language as machine commands is impressive by itself. I would suggest that devices sitting on your counter may not be solely capable of doing all those different levels of magic. They may need to send the verbal noises emanating from humans to the mothership for interpretation. This fact is going to mean that collection of your utterings in the neighborhood of a device such as Alexa is inevitable.

This architecture of having the work of conversion and interpretation of spoken commands done via transmission to business servers is documented in a number of sources, including the Infosec Institute which specializes in providing training and resources to the computer security community.[4]

9.3 General Problems of Home Networks

Home networks share many of the problems that large corporate networks have. However, the problems with small office networks and regular family home networks are only going to be exacerbated because of even fewer resources than is typical for larger business networks. Large corporate networks, for example, are going to have dedicated appliances for network security and for things such as firewalls. Having those kinds of tools on a home network are going to be the exception rather than the rule.

So, it's going to be even more critical that home users have active and updated firewalls and anti-virus on each of the appliances that they use for interacting with the internet. Additionally, it is going to be even more critical that users of these various devices engage in basic good practices such as, in addition to virus and firewall protection, not surfing with accounts that have administrative rights.

The preponderance of the Internet is infected with viruses. Not having your system protected means that you are very likely to catch a virus in your wanderings. If you are surfing the Internet and the account you are using does not have administrative rights, you are way less likely to have your device get the virus. Most programs require installation, and not having administrative rights means that a random virus picked up on web surfing adventures is way less likely to get installed in your computer.

A variation of this is a major element of what protects Linux and Unix systems. To install pretty much any piece of software, you have to get permission of the central part of the operating system, the kernel, to do any software installation. Not surfing with an administrative account pushes your security up toward the levels enjoyed by Linux/Unix systems, because you have to explicitly stop and give the administrative password in order to get the software installed and functioning.

Recommendation: Always create a secondary account that does not have administrative privileges to surf and do regular daily tasks with. One of the big problems with using accounts having administrative privileges is you can be the victim of a drive-by download and installation. Using non-administrative accounts prevents that kind of drive-by downloading and installation from happening.

9.4 The Internet of Things

We've been nibbling the edges of the issue of the Internet of Things. The Internet of Things consists of those devices that have the ability to be accessed and updated via the Internet, which includes a wide variety of device types. Things typically hooked up to the Internet of things typically include, smart TVs, baby or pet monitors, smart appliances like coffee makers, smart environmental control systems, and home security systems (the last two categories can and do often overlap).

Since we are talking about a wide variety of different devices, we are also talking about a wide variety of different kinds of threats. Let's look, in general terms, at the categories outlined and see if we can discover a pattern of issues in each.

However, before going there, let's outline some problems that are endemic to pretty much all the devices that fit into this category. First, there is a wide range of quality of components used in the appliances that are part of the Internet of Things. Many of the devices will have components including a processor and programming that are ill-protected, if at all, from providing administrative access because they are just not adequately secured with a reasonable password. It's is not uncommon to have passwords of "password" or the ever-lovely "12345." Why do smart devices come with passwords like this? Remember, often, they are run by programs that need, themselves, to be periodically updated. Further, finding a smart device at an IP address without a reasonable password, many attackers will not hesitate to attack.

The difficulty comes from many web-enabled devices having precisely the security problems just discussed. You may rightly ask how that problem is a problem for me, the person purchasing something that can be monitored or accessed over the web. Well, in addition to having someone violate your privacy without your permission, you have the problem of your device being used as an element of a denial of service attack (DOS or DDOS).

Denial of service attacks are where someone pretends to connect with your computer. In making a false attempt, some time is taken by your computer or router to process the pro-offered connection request. Enough of these requests can mire a device not permitting it to do anything else. DOS is where one person is responsible for the attack, DDOS is where multiple people or computers infected with a virus coordinate an attack.

The nasty consequence of DOS attacks is that participants, even unwitting ones, can end up losing access to the Internet — usually temporarily. ISPs will not hesitate to block access to individuals who do not keep their computers and networks relatively clean of viruses and malware that do things such as making a bunch of false requests for connection.

IoT devices have been notorious for the ease in which they can get incorporated into DOS or DDOS attacks. In fact, some of the most notorious major DOS attacks have featured compromised devices on the Internet of Things. One, a few years ago, managed to cripple a major portion of the Internet in the United States, albeit for a short time.

9.5 Baby Monitors, Pet Monitors, and Nanny-cams

The security issues associated with tech related to baby monitors have become nearly legendary. I would bet real money that you have probably heard of some of these issues. Some of the more famous ones include multiple issues where the web-enabled baby cam was taken control of by hackers who proceeded, in some very frightening ways, to talk to the child being monitored.[5,6]

This is one area where you don't want to go cheap. While more expensive systems are not a guarantee of security, inexpensive components developed and deployed with little or no concern regarding security will really raise the probability that easily compromisable components have been employed.

Recommendation: Frankly, this is an area where you really want to do your due diligence and research the quality of the components used in the device you have purchased. Doing this is no guarantee, but reviews, both at the point of sale and on the web in general, will definitely screen out some of the worse contenders for your dollar.

9.6 Home Security Systems

In doing research for this chapter, it surprised me to find out that some actually will ask the question: "Can a home security system be hacked?" If it is web-enabled, the answer is a resounding, "yes!"

Many of the issues raised with baby-cams will apply here. Do your due diligence. Research the product. Don't go cheap. Avoid the temptation to "roll your own."

Why do I suggest that you want to avoid building your web-enabled home security system from components that you assemble yourselves? First, many will do precisely that with the expectation of saving themselves money, and in doing so, they will often not thoroughly assess the quality, security, and interoperability of their components, and in doing so, problems like insecure passwords will often be endemic to the systems built. Second, the issue of interoperability of components can create additional security gaps in cobbled-together systems.

There is another level at which you do not want to "roll your own" home security system. Usually, most will want their home security system monitored. Home security systems are what tech types would call an "input–output" system. You can look in on them, but in sending messages to you or to your monitoring agent, they also perform output. Home rolled systems have had the problem of experiencing incompatibilities among components which can lead them to call for help when there isn't a problem. Depending on who is on the other end, one can rapidly find themselves in a situation where their system's cries for assistance rapidly get discounted. Again, we have the boy, the wolf, and a potentially disastrous situation.[7,8]

Recommendations: At the risk of being repetitive: due diligence! Research the system, its quality and reliability, and research the reputation of the monitoring firm. Pay particular issue to any information that you can find relating to known security issues and attacks against your home security appliance.

One common theme that needs to be repeated with all smart home appliances and components is the importance of keeping them updated. This applies not only to smart TVs but to any appliance hooked up to the Internet of Things.

9.7 Smart Environmental Control Systems

These systems share many of the same potential problems and issues as home security systems. Components may differ, but the risks and the preventatives are the same. Choosing quality systems with good components that you keep updated are essential elements in ensuring the most security and what shred of privacy that you may be able to continue to hold onto in the modern world.

9.8 Conclusion

Home networks bring a whole new range of issues having the potential for impacting privacy and security. They bring a wealth of entertainment, information, and convenience. However, the price for all these benefits, in addition to the expense of acquiring all the devices enabling all this fun, a cost in increased risk to both security and privacy have to be paid.

A major factor in protecting ourselves is to be careful in how we implement our home networks. Make sure that you encrypt your home wireless and you use tools like WPA3 in setting up the security of your home wireless network.

Another factor that you should consider in protecting yourself and your security is how you choose the elements of your home network. Make sure that you choose quality systems for things such as home security systems and smart environmental control systems. Going cheap and rolling your own home security system has been shown to increase security exposure for individuals. Keep your system updated.

Smart TVs are singular in the increased levels of privacy and security risk that they bring. It really is very much important to keep your Smart TVs updated. However, many continue to be uncomfortable with the fact that TV manufacturers have been shown, in multiple situations, to be using TV cameras and microphones to monitor their customers. Many recommend that it is better, from a security standpoint, to rely on smart streaming systems like Apple TV, Roku, Firestick, or Chromecast.

Be very cautious in the use of digital personal assistants. Their fundamental function and the ways in which they process information are inherently prone to problems with security and privacy. Use them if you wish, but be very aware of their potential for multiplying the availability of information about you and your life.

Chapter Notes

1. "Smart TV Security Concerns and 3 Ways to Stay Safe," Amber Mac, Bitdefender, accessed 10 December 2022, https://www.bitdefender.com/blog/hotforsecurity/smart-tv-security-concerns-3-ways-to-stay-safe

2. "New Vulnerabilities in Smart TVs Could Allow Hackers to Spy on Users, McAfee, accessed 10 December 2022, https://www.mcafee.com/blogs/internet-security/vestel-firmware/"

3. "Amazon's Alexa Collects More of Your Data Than Any Other Smart Assistant," Jason Cohen, PC Magazine, accessed 10 December 2022, https://www.pcmag.com/news/amazons-alexa-collects-more-of-your-data-than-any-other-smart-assistant

4. "The Virtual Personal Assistant and Its Security Issues," Ravi Das, INFOSEC Institute, accessed 10 December 2022, https://resources.infosecinstitute.com/topic/virtual-personal-assistant-security-issues/

5. "A hacker can take control of this baby monitor featured on Amazon," Tekdeeps.com, accessed 9 November 2022, https://tekdeeps.com/a-hacker-can-take-control-of-this-baby-monitor-featured-on-amazon/

6. "Baby monitor hacked, spies on Texas child," Chenda Ngak, CBS News, accessed 9 November 2022, https://www.cbsnews.com/news/baby-monitor-hacked-spies-on-texas-child/

7. "Security Challenges in the Internet of Things," Demitar Kostadinov, Infosec Institute, accessed 9 November 2022, https://resources.infosecinstitute.com/topic/security-challenges-in-the-internet-of-things-iot/

8. "Internet of Things Security Risks," Avast.com, accessed 9 November 2022, https://www.avast.com/c-iot-security-risks

CHAPTER

10

The Cloud

I frequently hear people stating how they do not trust the cloud. Alternatively, I hear them asking the doubt-filled question of "should I use the cloud?"

For me, the answer is almost invariably "yes." However, one really has to know what you are getting yourself into. There are different kinds of business models being used by storage providers. These different business models will impact how likely the provider is to try to snoop through your data. It will also impact how likely it is that you might lose large amounts of data because the service provider has been shut down by the government for gross copyright violations.

Also, we need to be clear about one basic thing. With "the cloud," you are not just looking at the issue of storage. There are lots of different kinds of computing services that you can get access to through this amorphous entity that we have named "the cloud." Any survey of what cloud computing is needs to at least offer a passing nod to those kinds of computing services. First, however, we'll focus on the issue of storing your stuff on somebody else's computer.

10.1 Cloud Storage Business Models and How They Affect You

Risking being trite, free isn't free. Microsoft, Google, and Amazon all offer free storage. Initially, we'll focus on the "free" storage coming from Microsoft and Google.

Google is the pre-eminent provider of free computer services that most of us deal with on a daily basis. Many of us forget that Google is not in the business of providing free services. Google is an advertising company. It carefully follows your surfing and searching habits with the intent of being able to more effectively target advertising to you. The same thing is true of the free software productivity suite that it provides. It is also true of the free cloud storage that it provides. It should not be a surprise to discover that Google goes through the files that are stored on it with the intent of discovering the kinds of things that you are interested in so that advertising can be more effectively targeted.

The snooping that Google does seems to be all aggregated snooping. This means that what it finds it does not seem to associate with an individual. Rather, it associates it with aggregated stuff such as your age, education, financial status, and so on. These tidbits can then be provided to advertisers to help them more effectively aim their ad campaigns. Google may not be providing information to advertisers that can be directly identified with you, but the aggregated information that they provide can be used to more effectively target you with ads that more completely reflect your interests.

As mentioned elsewhere in this book, Google is fundamentally changing the equation regarding how people are tracked. Google is replacing cookies with something they call Federated Learning of Cohorts. Instead of using a cookie to track where you go and what you do, Google is deploying watching programs to track your activities through your adventures on the web. Furthermore, Google is intending on not permitting the use of cookies in Chrome. Instead, their intent is allowing tracking to occur outside the individual web application. The cookies currently used to track you across multiple sites are commonly called third party cookies.

This limitation on third party cookies will have interesting effects. It will lessen the number of eyes watching you — generally a good thing. However, it will also strengthen Google's ability to peer into the daily aspects of how and what you are thinking as one employs the web.

So, you need to ask yourself a question: are you comfortable with someone poking through your personal information (even in an aggregated fashion) in order for you to be able to get free (or substantially discounted) cloud storage? Some elect to answer no across the board: I'll continue to use my SD card, my jump drive, or my portable hard drive, thank you very much.

For many, continued use of detachable storage is not an effective or reasonable answer. There are many reasons to use cloud storage. One of the biggest is the fact that, no matter how good, your storage device is eventually going to fail and if you don't have multiple copies of your data someplace, you will be lacking in fortune, and this lack of fortune could be very, very costly. Hard drives fail. Jump drives fail. The devices we use for storage on a daily basis have a limited number of times in which they can be written to or read from. In many ways, every time we use a device, we're making a bet that the equipment won't croak on us that time.

Personally, I prefer to put my risks in the RAID 5 storage, clustering, load balancing, and other server-based techniques employed by service companies to ensure high availability of data. Server and industrial grade storage devices are inherently designed to be far more reliable than the consumer grade devices most of us end up purchasing to hold our family photos. When combined with advanced server redundancy and backup technologies, the answer is clear. It's foolish to rely on a jump drive when one can invest in far more effective levels of data availability — data availability inherent in cloud storage.

Many have raised the issue of the problem with cloud storage coming from the fact that one only has access to the data if one has access to the internet. So, the narrative goes, if you don't have access to the internet, you are in trouble regarding access to your information. However, let's remember one very important fact: this is no longer the 1990s and internet availability is pretty much ubiquitous. For many, computational devices don't really have an existence outside of their ability to connect to the Internet. Every little restaurant, coffee shop and publicly oriented service business advertises free internet. Open your laptop and check wireless availability. At any one time you'll find bunches of wireless signals available — many of them foolishly left open and unencrypted. While I would never advocate attempting to access the internet via somebody's unknown and unprotected wireless connection (it's not safe and it's not ethical), the point is simply this: objecting to the use of the cloud based on the availability of the internet is now simply ridiculous. We are constantly being bathed in the glow of electromagnetic radiation representing the transport of information over the internet.

One aspect of the general ubiquity of net access that one should consider is this: free and public wireless, as discussed in the last chapter, which is a staple part of going to a restaurant or to a coffee shop, is generally not safe access. When accessing your potentially critical personal information on the cloud via an unsafe public wireless, you are heaping danger on top of danger.

However, let's get back to our discussion of cloud services and how they should be used. Let's consider Amazon's cloud storage service. This business model is straightforward. Amazon provides storage services so that they can more effectively sell you content. They want to sell you music, movies and books.

If having free or cheap availability of the storage of the stuff you've purchased makes a difference in their ability to sell you stuff, then it is a smart investment on their part to sell you the space to store the things you've bought from them. Amazon also has a paid service.

10.2 Play for Pay Services

There are lots of services where you can purchase storage to be accessed via the web. We're not going to do any sort of survey of these services. However, some of the more common ones include Carbonite, Sugar Sync, Dropbox, Microsoft One-Drive, Spider Oak, Apple's iCloud, Mediafire, Mega, Google Drive, Ubuntu One, and iDrive.

The magic question is this: which of these services should you choose? The answer is simple, but it not necessarily straightforward. Choose the one with the features that best meets your needs. Some do an excellent job of encryption and implement solid SSL security ensuring your data is securely at rest on their machines and in transit to and from their servers. Some have a strong user interface that makes it easy to synchronize your information between the various devices that you have hooked up. Some specialize in providing business services. Others provide large amounts of free storage. Some are relatively fast. Some provide lots of free storage but are slow to use. Then, there is the issue of price.

Start by reading reviews. Lifehacker, CNet, and PC Magazine have all done reviews of storage services. Since making a specific recommendation as to a particular service is not what this book is about, I'll let you read reviews and come to your own conclusions.

However, I will note this. At this writing, I am in the process of changing cloud service providers. Originally, I chose the one I'm using because it had (note the tense) a clean interface that was very much like the file browsing interface on my primary operating system and because the synchronization was seamless and reliable. Since then, they have abandoned their clean user interface, and their synchronization has gotten spotty. In addition, their software insists on using a big chunk of a large hard drive that I have on one of my systems to download the information that they are supposed to be backing up for me. All in all, it is time to quit.

You may also want to read a sampling of news articles pertaining to different providers. Sometimes, news articles will provide a perspective on a service that you might not otherwise get through a simple technical and operational review.

Even though it has been mentioned obliquely, the issue of security of the different services remains. Even with the ones that have a reputation for security, one might have concerns that providers can go through sensitive personal and business information.

There are a number of remedies that are available to users who have concerns about issues of snooping. First, one can implement different encryption services that will allow the encryption of different files and folders before uploading them to a designated synchronization folder. Secondly, one can make sure that the information that is transported gets moved using VPN services. Combining these two approaches will help to make sure that your data is relatively secure at rest and while being transported.

To summarize, here are some basic issues that you will probably want to consider before selecting a service:

- How much free storage will one get (important to an initial evaluation of the quality and effectiveness of the service). Personally, while I insist on reading reviews of products and services to get an initial "lay of the land," I am going to make my own assessment of the service and how well I like how it works.
- What is the cost of the service? If you are a heavy user of the cloud, you will rapidly exhaust the storage capacity of the service, particularly if you are synchronizing multiple devices.
- Does the service allow you to perform automatic synchronization among the various computers that you use? Can you access files stored natively on another computer? This should be a given, but it isn't. It is, unfortunately, something that you need to check on in examining the different services that are available.
- Does the service allow you to choose which folders are synchronized? You may want to only synchronize some folders and having some control over what folders get synchronized is pretty helpful.
- Does the service have an effective user interface? There will be some variability in user preferences. I happen to prefer something with the clean user interface similar to my regular computer browser. My provider has gone with something that is more oriented to mobile devices, and it has made my experience at the computer a lot clunkier. This is one of the reasons why it is going bye-bye.
- Does the service prioritize security? Data transmission should be via SSL (HTTPS), and your data should be encrypted on storage. You can, of course, add to that level of security by doing your own encryption before uploading the data to the cloud.

- What is the reputation of the service? Some services have been found to be simple repositories for copyright infringement. There was the notorious case of Megaupload. In that case, the company was accused of systematically uploading and hosting copyrighted material. The case is still being fought in the courts. In the meantime, thousands of users got caught in this web of legalities and found their data had been frozen when an international legal task force descended on the service provider. Doing some web searches of the service will often provide pretty direct hints of these kinds of activities. None of the services mentioned in this discussion seem to be the next Megaupload. Oh, by the way, there is a new iteration of Megaupload residing in New Zealand.
- Does the service focus most heavily on the "free" nature of its service? If so, you may want to shy away from these services, particularly if the privacy of your data is of first importance to you.
- How fast is the service? Some services will store data for you, but you may find that their architectures are not designed to provide responsiveness. You don't want to be waiting for that business presentation proposal to load while a bunch of clients are waiting. Reading reviews will help with this issue, but you will need to really get the best information by simply testing the service yourself.

10.3 Other Types of Cloud Services

Storage is not the only thing that is moving onto the cloud. Software is increasingly becoming "hosted" software with more manufactures adopting a "pay as you go" stance for getting access to their products. Some of the services doing this include Microsoft with the 365 services, Adobe with their Creative Suite, and Google with their Docs suite. Still other interesting services are provided by organizations such as Amazon. We also need to mention "Apps Stores" that are used by Apple and Google as ways of providing distribution of software for MAC OS/X and Android computers.

It's been no secret that for some time software distributors/developers have wanted to adopt a pay-as-you-go approach to selling software. Adobe has made a very strong foray into that arena with its "Creative Suite." Without making an inclusive list of all products involved in Creative Suite group, we are talking about a pretty extensive list of graphics design programs, including Photoshop, InDesign (a desktop publishing program), Adobe Flash (animation software), Dreamweaver (web design software), and several others.

What many individuals appreciate is the ability to start and stop these kinds of services when they are needed. Others are concerned about the ability to continue to have access to their information in light of not owning permanent access to the software that was used to create the documents.

The question most germane to this discussion remains, "are these types of software rental arrangements secure?" Looking at the issue of whether or not the company can look at your creations, we have to refer to the issues that have been discussed previously in this chapter. Certainly, companies have the ability to include intrusive code into the programs that they sell, whether or not they are cloud based. Inherently, this privacy concern is something that is pretty much implicit in all proprietary, closed-source software that one might be able to rent or purchase. So, it is not something that is more of an issue with cloud-based applications than it would be with any other purchased software. There is, of course, an exception to this assessment: one can always disconnect their computers from the internet to limit the opportunity for software to phone home.

As for other security concerns, one needs to attend to the issue as to whether or not the connection that one is using is encrypted or not (SSL?).

Microsoft has been moving toward cloud-based services for some time. That is the primary intent of the Microsoft Office 365 line of products. Microsoft has plans that begin, at this writing, at $5.00 per month. Functional capabilities are beyond the scope of this discussion. The same security issues that were presented regarding Adobe products would also be applicable to Microsoft products.

Many organizations are relying on the cloud for the distribution of software. For example, Apple makes extensive use of its App Store for the distribution of software to its PCs as well as for its other hardware products. Many fans of Apple products refer to the "Apple Ecology." The Apple App Store is an integral element of that ecology and many Apple software products are distributed through that venue.

Regarding security and privacy issues, users should be aware of certain approaches that Apple has taken toward privacy and security. First, it is Apple's stance that their products do not get viruses. While it is definitely true that Apple products get viruses a lot less frequently than do other systems, they are most definitely not immune to viruses. The resistance of Apple products to viruses is a great thing for end users. Apple's stance on their products not getting viruses is not always so good. For example, I've seen what can happen to an iPhone user when caught between Apple and the cell service provider over a security problem that may or may not have been Apple's issue. Apple's behavior to the end user was not helpful.

The second issue needing to be re-examined is the fact Apple has been fairly aggressive about writing software that benefits Apple first and that may be considered harmful to the purchaser of the product. Several years ago, Apple sought to patent a product that organizations such as the Electronic Frontier Foundation have since come to describe as "Traitorware." These programs might be used to determine if you are doing something with the device that Apple doesn't like, such as attempting to root a device to install a new operating system. Upon detecting this attempt, one reported behavior of "Traitorware" is taking your picture and sending the GPS data back to Apple so you can be tracked. To be balanced, this kind of tool can be used to help track thieves.[1]

This may or may not be problematical for individual users. Many find jailbreaking outside their interests. Some consider the ability to trace a stolen device to be a real bonus. Users will have to weigh the advantages and disadvantages of computer and device manufacturers being able to peer into their devices.

As an example, consider the Apple product Siri. This kind of personal assistant has fairly extensive capabilities of being able to take currently spoken commands over devices such as iPhones and respond to them fairly seamlessly and directly. What many may not consider is that those commands are not necessarily processed locally. As technology currently stands, the capacity of artificial intelligence processing to handle voice commands in as sophisticated a fashion as one might experience with Siri needs additional processing not always present in the small portable devices. So, the commands are often routed to a remote server for interpretation — something to be considered if you are a privacy fanatic.

10.4 Concluding Issues Regarding Cloud Computing

Cloud computing is going to get more extensive, not less. Increasingly, providers such as Amazon package entire sets of cloud-based services that developers can employ in the development of increasingly sophisticated sets of software services. New software development environments, such as Ruby on Rails, specialize in the packaging of services so that applications can talk to clients and servers for extraction and manipulation of data in new and interesting ways. The power of tools like Ruby and Ruby on Rails means that sophisticated packaging of cloud services for end users will become easier and much more prevalent.

End users need to be ready for the age of the cloud — with everything that means for privacy and security. New and unique applications that come from cloud processing and resources will push more and more developers to use programming environments to create the kinds of applications that users will find interesting. The cloud is here.

Chapter Notes

1. "Steve Jobs is watching you: Apple seeking to patent spyware," Julie Samuels, Electronic Frontier Foundation, accessed 12 December 2022, https://www.eff.org/deeplinks/2010/08/steve-jobs-watching-you-apple-seeking-patent-0

11

Acquiring Software

11.1 Software Sources

There are a number of different options for acquiring software. You can download free software (using different types of software and licensing). You can pay monthly fees to acquire access to online software systems. You can purchase software from different kinds of stores or directly from manufacturers. Some kinds of systems have software stores where software for that platform be downloaded. Each option has strengths, weaknesses, and issues you should be aware of.

11.2 The Ins and Outs of Free Software

When looking at free software, the issue of licensing immediately becomes important because it impacts the usefulness and safety of your software. Accordingly, there are several different licenses of which you should be aware. These different licenses include: freeware, shareware, open source software, and General Public Licenses.

11.2.1 Freeware

Free software, commonly referred to as freeware, wildly varies in quality and safety. Many businesses have a direct and complete prohibition against the installation of free software on their business computers. This simply stems from the fact that it is not uncommon for virus-writers to put bad code into these kinds of programs. Of course, you will recognize this as a type of malware known as a "Trojan Horse."

Additionally, the thing that might be incorporated into your free software might not be directly a virus. There are other forms of malware that can do you various levels of harm. One of the more common variants is something called spyware — a little concoction that specializes in tracking your activities and reporting back to a master. This can be through the embedding of tracking cookies on your system or it can be something as completely nasty as putting a keystroke tracker into the program you have downloaded.

So, how does one guard against these various forms of nastiness being part of the free software you are downloading? First, research the software. What are different sources saying about the software in question? You can try general web searches to see what the hivemind of the Internet is saying about the program in question. There is a discussion forum on Reddit that is dedicated to discussing the pros and cons of different kinds of freeware. It is called, you guessed it, freeware.

Beyond validating the software, you should also strongly consider evaluating the software download site. As of this writing, some of the better web locations for software downloads include:

- The website of the developer (who'd'a thought this?)
- Ninite.com
- Softpedia.com
- MajorGeeks.com
- Downloadcrew.com
- Filehippo.com
- Filehorse.com
- Filepuma.com
- Snapfiles.com
- Cnet.com

There are certainly a lot more sources than these, and some of the other sources will be discussed from a different perspective later in this document. I'll discuss some of the more popular download sites.

Ninite.com: This is an excellent resource for individuals wanting to install common utilities like browsers, anti-virus programs, online storage apps, media players, and popular office programs (not including MS-Office). This is not a comprehensive list, and I recommend that you visit this site to get a full idea of the programs that are available through this excellent resource. You can identify a whole range of different applications that can be installed on your computer at this location. A warning at this point: this site is primarily oriented toward Windows users.

MajorGeeks.com: This is another excellent tool for people wanting to find a broad range of resources for doing various tasks on your computer. Again, you really should make it a point of visiting this site and exploring it to get a better idea of the kinds of things that it can do for you. This site is not as automated as is Ninite, but you can still get a very broad array of software from it.

Filehippo.com: I've generally had good experiences in downloading free utilities from this site. It is definitely worth a look.

CNET.com: CNET provides a pretty broad array of free software of various types. They have information regarding the type of licenses: freeware, open source, shareware, and so on. They also provide reviews of the software. Access the CNET.com download site with the following URL: https://download.cnet.com/.

General comments: When using download sites for freeware, be aware of the fact that many of them pay the bills through advertising. Sometimes, the basic site design can be kind of click-bait-y. One generally needs to be mildly cautions to not click on a link that takes you to some Internet netherworld.

11.2.2 Shareware

The basic idea of these kinds of software is "try it and if you like it, buy it." That bargain manifests itself in several ways. You may get the full version of the software on downloading it only to find out that the software has been "time-bombed." This means that you have a set number of days to test the software before it shuts down, and you really can't fool these programs because they embed themselves very deeply into a basic mechanism in your computer called the Registry, which identifies your computer and the various resources installed on it. Trying to do something like uninstalling and reinstalling software to get some additional days of free usage just doesn't work. You are pretty much forced to buy the software if you want to continue to use it.

The other version of shareware is often called "cripple ware." You get just enough of the functionality of the software to whet your appetite. Getting the full features requires purchase.

When using software download sites, be careful to read and do the following things:

- The review of the software, if they are available.
- The type of license and what is provided in the download (freeware, shareware).

- The precise information provided in check-boxes if you are having to provide information to enable the download, such as your operating system (many providers will sneak in nasty options such as browser customizations like search bars for your browser that can be a pain to get rid of).
- Use a throwaway email address if they ask you for an address. Some options include 10 Minute Mail, MailDrop, EmailOnDeck, ThrowAwayMail, and Guerilla Mail.

11.2.3 Open Source Software

Open source software is just as advertised. The source code statements created by a programmer used to make the software has been published, somewhere, at some time. People in the computer industry often have very high regard for open source software in that the ability to look at the statements making up the programs often means that more flaws that might otherwise impact the user have been detected by a much larger array of examining eyes.

Open source software has been a major part of the building of the computer industry. Much of the Internet was built on open source software such as the Unix operating system.

Just because it is open source does not mean that it is not owned by someone and that they are not exercising licensing restrictions on the availability of their product. For example, some companies will publish an open source product for free and will also publish a premium product — for a premium price. Make sure that you understand which one you are getting.

At this point, it might be interesting to note that very popular proprietary operating system, Mac OS-X, is an example of something that was originally derived from open source software — in this case, Berkeley System Development Unix.

11.2.4 General Public Licenses

General Public Licenses are a really different approach to computer software. When looking up the idea of a General Public License, many people will only read the introductory descriptive material that these licenses, according to Wikipedia, provide end users "the freedom to run, study, share, and modify software." What is really important in General Public Licenses is the fact that they institute something called "copylefting."

This issue of copylefting is really more important to end users than many realize. The concept, originating with programmer Richard Stallman, places an additional requirement on software licenses that any program modified from a copylefted product must also remain freely available.

The concept behind this idea of copylefting is to keep software free of controls because software is, in the view of copylefting advocates, an idea and ideas should not be controlled.

If you are looking at acquiring some GPL software, there are some organizations you should be aware of. Two closely related organizations that do a lot regarding free software such as maintaining indices of what's available include: The Free Software Foundation and GNU. Their websites are:

- For the Free Software Foundation: www.fsf.org
- For GNU: www.gnu.org

Some very important and useful pieces of software include:

- GIMP (GNU Image Manipulation Program): a completely free alternative to Photoshop.
- Linux: a free operating system that is very much like the operating system that built the Internet, Unix.
- Open Office: a free alternative to the industry dominant Microsoft Office. This program permits the creation and editing of MS-Office files.
- Libre Office: another free alternative to MS-Office. This software is a "fork" of Open Office.
- MySQL: while this is not a GPL program, it is open source, and it is considered to be a very effective alternative to commercial database products. Personal and enterprise editions are available.

One term that was used in the previous descriptive materials was the word "fork." A fork of something occurs in open source and GPL environments when a different group takes up the maintenance of a particular piece of software and may decide to add additional features to the program.

The take-away that I would like you to have from all of this is that there is a very large and thriving community of people making and maintaining both GPL and open source software. Instead of paying a high price for a commercial product, you might be able to find something with fewer bugs that does just as good a job — for free.

11.3 Free Anti-virus

It has been my experience that many people first get into the wonderful world of free software because of anti-virus software. They get their new, gnarly laptop and, thinking of nothing else, they merrily use it until something dreadful happens: the anti-virus that was packaged with their shiny toy has expired. Whatever caused their realization, they know they need anti-virus, and the costs of commercial anti-virus software seems too high. So, they look for cheap or free anti-virus. In this process of exploration, folks like me often get asked my opinion regarding many of the free packages they encounter.

So, what do I think of free anti-virus?

Generally, my experience has been good. Over the years, I have used, in no particular order of preference: AVG, AVAST, AVIRA, Bitdefender, Panda, and Malware Bytes. There are most certainly quite a few others, but the ones listed are the ones I have personal experience with.

As stated, my experience has been good, and I would recommend any of these to users who have, through whatever circumstance, been put into a situation where they need anti-virus right now and they can't or won't afford to buy a commercial product.

You see, the manufactures are working with a variation of the old shareware, try-it and buy-it model. It is to their benefit to make sure that you don't get a virus so that you will upgrade to the niftier commercial product.

In my experience, some virus developers have actually designed viruses to attack mainline commercial anti-virus such as McAfee, Symantec, and Norton that come bundled with new computer purchases. When this is done, smaller manufacturers or those that don't bundle with new purchases have frequently been ignored by virus developers. This isn't always true, but it has been true enough over the years to make it notable.

So, if you need anti-virus, free can often do the job very nicely for you.

11.4 Software and App Stores

When I am looking for software, one of the first places that I go to are the various software stores that are associated with the various different platforms I use. There are excellent App Stores maintained by Apple, Microsoft, Android, and by several of the different communities associated with different versions of Linux.

One of the nice things about App Stores is the fact that the organization will effectively curate the selections available for you to use. This will generally include compatibility, but it can also include issues such as privacy and security. Sometimes the overall quality of the software will also be assessed.

Of the various different App Stores that are out there, I have encountered the most positive reports regarding the Apple App Store. Working with what many would call a "closed garden" approach to software development, Apple has done a very strong job in ensuring that the programs in their marketplace are compatible with their platform and are virus free.

In contrast, one has to exercise pretty careful with programs in the Android App Store. Before downloading and installing a particular program, make sure that you diligently find out the experience of other users who have used that software. In fact, Norton Security, a well-known provider of anti-virus and security software, has stated that hundreds of programs in the Google Play Store (Android App Store) are in fact malicious programs that have been disguised as applications.[1]

Microsoft has an App Store that seems pretty well rounded. You can buy many common apps and you can also get open source alternatives.

Communities like those involved in the support of some different Linux varieties also have App Stores. This includes the communities supporting Ubuntu and Mint Linux.

In general, I would recommend that the first stop that a person makes in looking for software would be the App Store pertaining to that particular application. Failing to find an application in the app store, I would look to utility download sites such as Major Geeks and Ninite. I would also look at the indices maintain by GNU or the Free Software Foundation.

Chapter Notes

1.   "Hundreds of malicious apps are showing up on the Google Play Store, disguised as legitimate applications," Norton Software, accessed 12 December 2022, https://us.norton.com/internetsecurity-emerging-threats-hundreds-of-android-apps-containing-dresscode-malware-hiding-in-google-play-store.html

12

Fun with Protocols

If you spend any time at all with computer security types, one of the things that you will literally constantly hear them muttering about is protocols. There are some very good reasons for this obsession, some of which the average user of computing devices, whether these devices be cell phones or desktop computers, should know about in order to protect themselves.

To get things started, let's define what protocols are. Here I am going to offer an informal definition that, I think, most computer and networking experts would be comfortable with. A protocol is a set of rules enabling communications between devices.

Computer communications and networks are literally driven by protocols. They exist in many ways, and often, they are layered, meaning that you will often have multiple protocols in operation when you are doing a task over the Internet, and these protocols will often be used by other protocols to help achieve essential parts of the communications task needing to be accomplished.

Let's look at some of the protocols that every user of computational tools should know about, with the idea of making sure that we understand some potential dangers that can exist in the fundamental rules that allow communication to take place.

12.1 DNS

DNS or Domain Name Services are a set of protocols that allow you to type in things like www.amazon.com and have computers understand the actual computer system to which the information is to be directed. One thing that you need to know is that things like www.amazon.com are nowhere near the native language spoken by networks to allow information to reach its destination. Closer to the level used by networks is something called an IP address. IP stands for Internetworking Protocol, and it is part of a suite of protocols called TCP/IP where TCP stands for Transmission Control Protocol.

In understanding TCP/IP, it should be noted that there are different versions of this protocol, where the most current version is something called IPV6 or IP Version 6. In explaining TCP/IP, I'll base the explanation on the simplest version of the protocol — which is the somewhat older variant of the set of rules — something called IPV4.

In understanding TCP/IP, you should know that each network accessible on the Internet has a unique identifying number. In the older version we are looking at, we have four integer values that range from 0 to 255. So, under this schema, a typical IP address would look like 142.251.215.228, which is the IP address that Google appears to be at. Some integer values are specially reserved for important network applications. Some, for example, are only intended to be used in local networks and can't be accessed externally across the Internet. These include IP addresses that start with 10, 172, and 192. Networking types will describe these addresses as non-routable because they can't be used in forming a route to get someplace on the Internet outside of the local network.

What is important for us to know in looking at DNS and its potential dangers is that the purpose of Domain Name Services is to convert from what we are comfortable with, such as www.google.com into the IP numbers that are actually used to find networks on the Internet. The unfortunate part of this process is, because DNS is so important for us to use in order to find things on the Internet, it is often a pathway by which we can be attacked. What I am hoping to provide you is an awareness of what those potential attacks might be so that you can do things to minimize the general level of risk you are taking when you use the Internet.

In order to understand how DNS does things like converting www.google.com into 142.251.215.228, we need to take another side step to look at some more terminology that is closely related to DNS and its basic function. Things like www.google.com are also referred to as URLs or Universal Resource Locators, and they are an essential part of another protocol, Hypertext Transfer Protocol, or HTTP. This is the HTTP that commonly appears in the URLs we use to go to Internet sites. So, if we wanted to go to Google, we would could type in www.google.com in the address bar of our browser, but with older browsers, we would have had to type in http://www.google.com.

All of these elements are part of the process of implementing hypertext documents that we use in our browsers. It's all part of those protocols that allow us to click on links on webpages to take us far away to the wondrous things pointed at by our web pages.

Frankly, in order to understand how DNS works and how HTTP works, you really should understand the parts of a URL. Many times, URLs will end with something like .com, or .edu, or.biz, or .tv or .us (sometimes referring to the country that the site is at). There are others, but this list gives you a flavor of what kinds of things are being referred to.

The elements just mentioned are what we refer to as the Top-Level Domain or Domain type. There are master servers scattered around the world that are specialized, for example, in the resolution of .com addresses. There are others that specialize in .edu addresses. What these servers have is an index of those URLs that have .com in them for the parts of the Internet that are commercial enterprises. These servers convert, for example, www.google.com to 142.251.215.228 so that the networking protocols can locate the world's most famous search engine.

So, each domain type (.biz, .edu, .com, and so on) will have separate servers that specialize in the resolution of that type of address. So far, so good, right?

Well, things are a bit more complicated than this. There are so many different places that we might want to go to that we really can't funnel all the business traffic to just one server that specializing in resolving .com traffic into IP addresses. There are copies of these indices scattered all across the Internet. Your ISP undoubtedly has copies of these embedded in its own network. Again, in referring to these copies, you will hear networking types mumble about Name Servers. Name Servers do the work of resolving these requests for .com addresses or .edu addresses into the physical network addresses that we need to get some specific place on the internet. They are also responsible for indexing something called MX or Mail Exchange records allowing email to find its way on the Internet.

Here is where things can get really homely. There is a complicated process by which these addresses propagate through the Internet, and one is really quite dependent on the accuracy and security of these indexes that tell us where everything is. Sometimes, cyber criminals will try to corrupt this process by putting bad entries into the DNS index. This attack will often direct you to a website that the criminal wants people to go to instead of the entry or entries that have been replaced on the DNS index.

This attack is commonly called DNS poisoning, and it is relatively difficult to detect — except there will be a certain percentage of the users of a network who are being redirected a website that is acting just a bit off. The website where you go instead of the website you want to go to — for the parts of the DNS table that have been changed — will often look right, but it won't completely act right. When logging in, you might get quickly tossed out of the website — even if you enter your username and password correctly. What the fake webpage, made up to look like the right one, might be doing is trying to get you to enter your credentials, username and password, for that page so that the site developers can pretend to be you at a later time.

So, the question is what can you do to prevent this kind of attack from being waged against you and your device? The answer is that you can do one of two things: you can change the DNS settings on your home or business router to go to a more secure DNS server, or alternatively, you can change the DNS settings on your device. In both instances, you will be altering your DNS settings to use a faster and more secure DNS service to do the resolution of names like Amazon.com into IP addresses.

How you will do that on your router is something that will change from router to router, but if you elect to go that route, you will find that the change of DNS server used by your router has by default enhanced the security of all the devices hooked up to your home network, whether they are PCs, TVs, or IoT components.

To start the process of changing your router's DNS, you will want to look up the name and model of the router that you happen to be using to access the Internet. You'll be doing a web search. So, you'll then want to combine the name and model information with the phrase, "change DNS" to do the web search necessary to find the steps that you will have to go through in order to make the change of DNS become part of your basic network settings. Assuming that you have gone through the steps that you find with care, the result should be a faster and safer network for everyone at your house.

The alternative action is to change the DNS server on individual devices. You can do this fairly straightforwardly on most devices. Apple has good reference materials on how to do this on the Mac. On windows, you will have to go to your Internet connection and choose properties. You'll then look for Internet Protocol Version 4 and click on the properties button. You then should be given a series of options, one of which is DNS server addresses. I would enter at least two addresses.

At this point, the discussion has emphasized that you should change your DNS server address, but the values that you should consider have not been specified. There are several relatively safe, secure, and fast alternatives that are available.

The first and probably the most famous of the safe DNS servers are those maintained by Google. Google has two servers — one at IP Address 8.8.8.8 and the other at 8.8.4.4. Just set your device to use those two addresses for DNS resolution and you are on your way.

There is another company that has been also doing the safe DNS thing for quite some time. It is an organization called Open DNS. They are very easily found on web searches. Their DNS server locations are not so snazzily named as the other services, but they do have lots of interesting articles on DNS and security, and what's more, they have detailed instructions on how to set up your DNS server on pretty much all of the popular devices, PCs, gaming consoles, and whatnot.

Still another company that does safe and secure DNS is an organization called Cloudflare. They are a really big deal on the Internet. One of the things that they do is to help organizations protect themselves against distributed denial of service (DDOS) attacks, where, for example, a bunch of computers infected with viruses all act together to keep pumping connection requests to one server, thereby consuming all of its time and resources. This effectively can shut a computer down from being able to respond to requests from legitimate users.

One of the nifty things provided by Cloudflare is a secure DNS service. You can access it when you set 1.1.1.1 as your DNS server.

There are still other services that are available. However, choosing one of these services is a solid choice and any one will do a lot to enhance your overall security.

Recommendation: Do the research and change the DNS of your router. This will do a lot to enhance your security — especially when combined with the use of VPNs or even The Onion Router.

Recommendation: I generally recommend the use of a secure DNS service. However, if you use a mobile device and travel with it frequently, it is even more important that you set your DNS to a common secure DNS service. Given the fact that you don't know what kind of DNS servers you are going to encounter as you move from network to network, you really do need to take that element out of the mix as a possible source of attacks against you and your device.

12.2 Problems with HTTP and HTTPS

In addition to the problems that can come from DNS being compromised, there are some other security exposures that are inherent with hyperlinks as we use them in navigating the Internet. To illustrate what those exposures are, we need to return to our discussion of the basic structure of URLs.

To do this, let's look at a hypothetical URL:

https://acomputer.adomain.com/subfolder/file.aspx

Let me emphasize: this is a hypothetical URL that will take you no place if you click on it.

As mentioned previously, .com is the Top-Level Domain or the Domain type, and what it does for you is it helps determine which index that you are going to use to resolve the URL and look up the equivalent IP address. So, in this example, adomain is the actual domain name. Looking further into the example, acomputer is what we will call a sub-domain. Usually, this will mean a particular part of the network or a specialized server that is dedicated to a special category of functionality. For example, on a commercial website like amazon, this might be a server dedicated to something like streaming home videos.

Where things get interesting, however, is to the right of the domain, where you see entries such as subfolder/file.aspx. Those elements are often requests to run or use specific files. These requests often cause the execution of specialized sequences of commands that are called scripts. These sequences of commands can either be helpful or they can be very, very malicious. For example, these files can include sequences that cause the launching of things such as malware.

Recommendation: Be very wary of clicking on links. As mentioned, these can be requests to launch things such as viruses to attack your system. I would suggest that you make it a habit to copy links and paste them into the URL address bar. This will allow you to look at the contents of the URL and see if there is something unseemly about the link such as a request to run a script.

Recommendation: Elsewhere in this document, you will find discussion of tools that limit the ability of scripts to run. Let me re-endorse the use of these kinds of tools at this point. Script blockers will stand between you and the propensity of many web pages to automatically load and run scripts. They will give you the opportunity to decide whether or not you want to allow that web page to execute that sequence of commands it has on that page. Get a script blocker and install it on your browser. They are often implemented as browser add-ons.

There are lots of different script blockers. To find one, you may want to go to the add-in store for your favorite browser. Some of the more common script blockers include NoScript, Script Block, and Disable JS (JS stands for Java Script — a common scripting language used in web pages).

Some browsers, like the Avast Secure Browser, will include script blocking features as a basic part of the capabilities of the browser. Other secure browsers include Brave, Firefox, and TOR. Of the mainline browsers (Chrome, Edge, Firefox, and Safari), Firefox is probably the most secure and privacy-oriented. However, TOR probably does the best job in protecting your privacy.

A disclaimer should be offered at this point. Limiting the ability of scripts to run is often going to impact the functionality of the web pages that you use on a daily basis. Hence, most script blockers are going to give you the option to decide whether or not you want to run scripts on a particular page. They'll allow you to choose whether or not you want to temporarily or permanently disable scripts on whatever page you happen to be on at a time. Consider the question carefully. Disabling scripting protection permanently on a page really relaxes your security. You should really consider as to whether or not you trust a particular page that much.

12.3 Conclusion

So, in this chapter, we considered the fact that, by their basic nature, some of the protocols, or rules for communication, that we use have some inherent flaws that, of necessity, have to be built-in to them in order to work. One such inherently flawed protocol is the Domain Name Service, that we use on a daily basis to convert names that mean something to us, like www.google.com, into something that networking systems can use to derive the IP address actually employed by the Internet to figure where things are located. Cybercriminals will often attack the indices converting names into addresses via an attack called DNS poisoning. To enhance your security, it is recommended that you consider the use of secure DNS services, particularly if you travel with your devices.

We also discussed the fact that there are flaws built-in to the names that we use to move across the hypertext implementing web documents. To stand less of a chance if having lists of commands run against your computer, it is recommended that you copy URLs manually instead of just blindly clicking on them.

Furthermore, it is recommended that you employ script blockers to enhance your security when surfing the web.

CHAPTER

13

The Future

There always is an element of arrogance when one tries to peer into the future. There's always some little thing that one regards as insignificant that turns out to be the most significant thing that ever was. As an example of this, consider Bill Gates' book, "The Road Ahead." It was written just before the Internet exploded on our awareness. Initial copies of the book, so I am told, did not discuss the Internet to any real degree. Just after the book was released, the Internet "happened." I'm told that there was a huge effort to gather up copies of the book that overlooked the Internet. A quick revision was done, and the updated version was soon released to the world. Rumor has it that a few of the old version escaped the dragnet and those are apparently quite valuable. So, even somebody as inured to technology as Gates can get it very wrong. The rest of us are pretty likely to overlook some pretty significant things too.

However, there are some trends that one can look at to see where some important things are heading. We can look at those and ask ourselves what the future might hold for some issues, which can be identified, relating to privacy and security.

13.1 You Need to be an Active Participant in Protecting Yourself

One thing that will probably never change in tech is the appearance of powerful new tools guaranteed to really shake things up. One such technology is quantum computing. Quantum computing is supposed to be able to solve previous unsolvable problems. A major area of interest within quantum computing is quantum encryption.

Quantum encryption is supposed to be unbreakable. The problem is those tools, even if they are implemented, will still be parts of systems, and one of the key parts of any system is going to be human beings. So, the security mavens who hail invincible systems built with unbreakable encryption have failed to factor in humans and their flaws. Too many times, security systems have been based on tools that were supposed to be unbreakable only to find that some flaw in the implementation meant that the tool was fundamentally weak. Additionally, with humans involved in systems, something will break down because humans are flawed, corruptible, lazy, and well, human.

Even in the best of circumstances, our humanity is a "flaw" that can be exploited by the predatory who know how to use our instincts against us. Most of us seem to have an instinct to help and be cooperative. Let the nice kid with the jump drive print out his resume because he has been out of work for two years...absolutely. The problem is that the jump drive has a virus on it. Help the phone tech out by letting him into our phone closet. The difficulty is that he isn't really a phone tech. He is trying to breach the physical and communications security of your business. The instinct to help will continue on in people and it will continue to be used against us by the sly and predatory. Security breaches will still occur.

If anything, security breaches will increase. Too many unsavory elements find there is too much money to be made in doing dishonest things over the Internet, and remember, the Internet was not designed with security in mind. Just this morning, as of this writing, I read about the latest episode of yet another appearance of malware that hijacks your files, encrypts them using a much more sophisticated form of encryption than is typical for these kinds of exploits, and that employs anonymizing tools such as the Onion Router to make it even more difficult to trace where the security breach is coming from. Victims of these computer crimes are then informed by the program that they have to pay in order to get access to their files back. These programs are generally collectively called ransomware. Just a few years ago, the initial appearance of these kinds of advanced malware were greeted with horror in the technical community. Now, these kinds of occurrences are almost greeted with a yawn. In fact, articles regularly appear in tech journals that outline the latest new versions of ransomware. In some journals, the articles recur on a weekly basis.[1]

One of the things that is interesting about these kinds of attack is the fact that they are often accompanied by a demand for payment in Bitcoin or some other cryptocurrency in order to get the key to allow the unlocking of the files. While there have been a few instances in which the encryption has been able to be broken so that people can get their files back, the sad part of this story is that there usually is way to get your files back without paying the ransom.

Government counsels that one should not pay. Most choose not to pay, but lots of money is still being made through these blackmail schemes. So, be alert and don't get infected.

Another reason why security breaches will increase comes from the fundamentally democratic nature of Internet technology. Previously, where one had to be a member of a small select few who knew how to break computer systems and write viruses, this kind of information is now plentifully available on the Internet. Just now, as an experiment, I entered the word "hacking" on Youtube. I got 3,540,000 results. Putting in the phrase "hacking Metasploit" (a well-known hacker's tool), I got 49,000 results. You get the picture. Anyone, with a moderate degree of interest can gain access to some fairly dangerous information.

You may have noticed that, at the beginning of this section, I used the title, "you need to be an active participant in protecting yourself." Being an active participant in your own protection will be so necessary because of the growing legions of people who want to hack. An active participant in protecting yourself means doing a whole bunch of things discussed in this book. To provide an incomplete review, it means:

- Have current anti-virus on your system.
- Have an active firewall on your system.
- Make sure that your software, both application and system, are current.
- Have anti-scripting software on your computer.
- Choose your browser carefully (not all browsers are created equal).
- Use a password manager.
- Choose strong passwords.
- Don't reuse passwords.
- Do not sign into systems using http.
- Use VPN software or services.
- Use the Onion Router.
- Be careful what kinds of sites that you go to.
- When you are surfing, don't get click happy.
- Read the dialog boxes that are presented to you.
- Watch the preview of the URLs of links that you might want to surf to (this can sometimes reveal a seamy site).
- Don't believe what they tell you (just because a site has an image that claims it is secure, don't count on it).
- Be careful what you put on public wireless (never put credit card information where it might be captured on public systems).
- Don't shop with your bank card — shop with credit cards or paying services like Paypal instead.

Yes, there are wolves in the cybernetic forest, but you can make it harder for them by making them work for it and by making it not so profitable to try to create problems for you.

Despite all the best intentions, something can get through your defenses. Sometimes even smart people have critical personal information stolen from them. What it comes to is this: if they want you bad enough, they will find a way to get you. So, what do you do?

My basic answer to that is to insure yourself. Have some sort of identity theft service that can help you protect and recover your identity should someone try to steal or try to buy things in your name but for their benefit. I've had more than one brush with attempts to steal my identity or levy charges against me that were not mine. For example, I've had to deal with these instances:

- A major national financial institution having a data breach that exposed the personal financial records of its customers. This could have easily led to individuals having their identities stolen.
- Someone using my credit card to buy an expensive wedding at a wedding chapel in Las Vegas. At the time, I was happily (still am) married, and at that point, I had never been to Las Vegas. It's a good thing that my wife has a good sense of humor.
- Credit card information that I have used with more than one store have been breached and exposed over the years with the result that my credit card could have been stolen. I have had to cancel credit cards as a result of these events.

There are several different providers of ID Theft protection. The major credit service companies will provide these kinds of services. Additionally, there are companies such as Lifelock and Trusted-ID that will provide identity protection. There are undoubtedly others, but these are a good starting point in your research.

However, you do not have to use identity theft protection services to enhance your level of safety and security. Remember, the starting theme for this chapter was your own active engagement in your own self-protection. So, you really need to make it a habit to take certain basic precautions to protect your identity like making it a habit to be actively engaged with your credit history. From my standpoint, this means:

- Regularly checking your credit. Remember, you are entitled to a free credit report each year.
- Locking your credit down. You only have to execute this with one of the three major credit bureaus.

Don't trust the companies that have been breached. Ironically, one of the major and more notorious breaches occurred a few years ago with the breach of one of the three major credit bureaus, Equifax.

Companies that have been breached will often try to spin it to have circumstances not reflect so badly on them. Often, they will let the notification process drift for a dangerously long time, leaving the poor customer's flap on their union suit blowing in the breeze. Aggressively demand the cancellation and the issuance of new identification cards should you get news that your information may have been breached. I know this is a hassle. Trust me, I've been through it and the inconvenience cost me some money, but I am confident that I did the right thing and that I significantly reduced the amount of the exposure I was vulnerable to.

Follow security news. I know that this can be depressing because it seems to consist of nothing more than a litany of the latest breaches and exploits, but it can also tip you off to the fact your data may have stolen. Most of the providers of anti-virus will have pretty decent blogs about what is going on in the security world.

Research the people that you might be considering doing business with. There are several ways in which you can approach this problem. One of the easiest is to use the Better Business Bureau. One of the advantages of checking the Better Business Bureau rating is the fact that you can verify that there are "real" people on the other end of the transaction — not just some ephemeral chunk of software code designed to soak in financial data such as credit card numbers as the first step in attempting to steal your identity. For me, membership in the BBB is irrelevant. The knowledge that the company exists someplace and has a decent record in handling customer feedback is what is seriously important.

Make sure that you go to the website "Have I Been Pwned?" at this URL: https://haveibeenpwned.com/.[2]

This website will let you know which accounts of yours have been involved in breaches. If you find that any of your accounts have been involved you can do the following:
- Log in and reset your password immediately if you still use the account.
- Delete the account otherwise.

In the category of "for what it is worth," the term "pwned" is something often seen in hacking or gaming communities. What it refers to is the idea that you have "owned" or dominated somebody. Why the "P?" Frankly, I just don't know.

Deleting accounts can be a royal pain. Some websites make it nearly impossible to get rid of old accounts. Simply not using an account does not remove its existence. Deleting it from your password manager does not impact what is still in place on the server that manages your account.

There is a simple reason for you to get rid of your old, dead accounts. They create an additional security exposure for you. They give potential fraudsters the ability to leverage their existence into a claim to be you. Don't give them that lever.

With the growing capabilities of AI, it will be getting easier to create entities such as "chatbots" to mimic the reactions of a human being and that will automate the extraction of sensitive information from unsuspecting users. For those following AI news, there has been discussion of a "bot" or automated software passing the Turing Test verifying that a program is actually intelligent. One of the implications of passing this test is that human beings are unable to tell whether the entity on the other end of the conversation is human or is a machine. With chatbots effectively acting like people, this capability of using software to mimic human behavior and information extraction means that automated software identity theft will soon be common.

Since we're on the subject of chat, one rule that I have is to never put critical personal and financial information out via a chat client. I don't care how much that a particular company claims to care about your privacy. Chat clients are notoriously insecure.

The point that I'm making at this point is this: if I can't verify a physical address, I don't do business with a company via the web. Admittedly, this refusal may cut out many potential businesses, but it does provide an additional level of reassurance as to who you are dealing with.

To restate, the organization you are thinking of interacting with does not have to be an accredited member of the Better Business Bureau. However, you can get an idea as to what kind of history that the organization has — do they have a record of creating and not answering customer complaints? If there is a track record, the BBB should be able to provide it for you.

Other ways exist allowing you to research the quality of a business. One that I have used is to check company web pages for potential industry professional memberships. Do they belong to the Amalgamated Widget Manufacturers of America (a hypothetical organization)? What is the reputation of the AWMA? That tidbit can be researched. I've used manufacturing organizations to validate companies and have successfully concluded business as a result.

Besides using tools like the BBB, one level of assurance that has proved useful to me is to execute payments via tools such as Paypal. I will do business with people simply because they have things configured to allow Paypal payments. Paypal and credit card companies are built around customer trust. A lack of customer trust means that Paypal and credit card companies can't exist. Paypal and credit card companies are ruthless in enforcing trust. I have had several businesses pay immediate attention when I filed a formal complaint through Paypal. Paypal immediately froze their accounts until the issue was resolved.

Credit card companies also have a business model that is based on the perception of integrity and safety by the customer. You will note that there is a distinct limit on the amount that a customer is liable for should a fraudulent transaction be applied to a customer account. Accordingly, the last thing that I would use for web-based transaction is a bank card.

In addition to the preceding methods of verification, you can also pay for background checks on companies and people on the web. One would also want to verify the organization doing the check. The levels of validation and verification can get pretty deep fairly quickly. It's worth it — especially if it helps to keep your identity from getting shredded.

In all of this, don't forget the role of the Public Key Infrastructure in helping to provide digital certificates and signatures for verifying someone is who they say they are. While digital certificates and digital signatures sometimes get compromised, they still remain our best method for validating that someone is who they claim to be. So, pay attention when your system complains about an expired or bad digital certificate. It is attempting to follow some very basic instructions to try to protect you and your computer.

Similar to your browser complaining about a website, you may get a message from your computer complaining at a software installation attempt. If you get a message complaining about an application's digital signature, attend to it. There are a number of reasons you might see a message warning you about an application you are trying to install. For example, the website creating the software might have an error or the program trying to be downloaded may have been modified. Either or both may be indicated by the existence of an error in a digital signature. While the situation may be completely innocent, I wouldn't risk it. Don't install the software, go someplace else to get it. If it is shareware, freeware, open source or General Public License software, it should easily be available in multiple locations.

So, there is a basic underlying theme: the hazards of computing are going to increase. Attacks will increase. Quantum cryptography and other quasi-magical computing solutions will not solve all our problems because you still have to deal with humans and human error. AI, despite its potential to help us, will also serve to make things more dangerous for us.

13.2 That Which We Do to Ourselves

Despite the seeming lack of control we have over technology, we also make choices impacting how tech affects us. As an example, consider home computerization — the trend toward having internal operations of our homes and businesses controlled and through the Internet. As noted, there are downsides to computerized control over home operations. That said, computerization is going to expand, despite anyone's squeamishness.

Presently, we have the early stages of tools that will become a common element of homes. Inside homes, we have Alexa and Siri as digital assistants allowing us to issue verbal commands for controlling devices and getting information. We also have the smart televisions on the IoT guiding our entertainment. Furthermore, we have the growing use of computerized environmental control systems, again implemented as elements of the Internet of the IoT. We also have home security systems on the IoT.

Do you see where this leads? We have the beginnings of the future. Many predict the development of the computerized home will be shaped by the home security gateway. Home security gateways will form the central hub for the computerized home. It will be the focal point by which the various elements of the Internet of Things will be controlled, including home security systems, home environmental controls, and the various smart appliances, like coffee makers, that will be controlled remotely.

Home security gateways have elements of AI built into them, permitting command of their operation by voice or gesture. The power and complexity of these systems continue to grow exponentially. This complexity growth means that security problems that we will have inflicted on ourselves will grow in a similar fashion.

Each externally controllable device multiplies potentially attackable security holes. Admittedly, the use of central control locus, like home security gateways, may limit the number of ways in which the various devices of our home can be abused, but the success of central control in limiting security holes will be determined by the gateway's programming.

Verbal control means that our gateways will always be open for abuse. How fast will our smart devices develop the ability to process spoken language requests directly and not send them for interpretation? Watching the evolution of tech, that is certainly quickly. However, will vendors design their equipment so that verbal requests will not be sent home for interpretation? I would suggest that manufacturers will opt to continue to watch us.

Additionally, one must consider the nature of what is being done by things such as home security gateways. They are to monitor and report the status of the devices they control, particularly security-oriented devices. Most users will have their home security gateway connected to a monitoring service.

Even with all the inherent scariness of this world, the movement to the use of tools like home security gateways will undoubtedly continue. So, the question now becomes this: what can I do to help protect my security and privacy?

13.3 Recommendations for Protecting Yourself in the World of the Home Security Gateway

- Change both the username and password for logging into the gateway. Remember, the poor passwords assigned by manufactures as discussed in our examination of the IoT. Hackers also tend to look for usernames like Administrator.
- When connecting a home security gateway to a wireless network, be especially careful to attend to the rules for security in wireless networks.
- Turn on automatic updates for your devices, if available. Attacks are often levied against non-updated systems.
- Don't use names for naming your gateway or wireless network that identify you in any way. Remember, if you can see their wireless networks, then they can most certainly see yours also.
- As with home security systems, avoid the impulse to "roll your own" and to use cheaper, poorly-secured components.

13.4 AI, the Not so Hidden Gremlin

Hiding in the background of all of this is the increasing use of AI technology. AI tech is a major element of home security and IoT gateways. AI allows the issuance of verbal commands for the command of the elements of our homes. However, AI also presents whole new pathways by which systems can be attacked. The problems exist at several different levels.

First, those who attack systems will increasingly use AI to attack systems. On the other side, security will increasingly have AI built into it allowing the prevention of security violations. However, given the growth of systems complexity, there would appear to be an advantage for those who attack systems.

Second, a whole new class of attacks have been based on the nature of AI and it involves training AI to do what it is supposed to do. AI learning may or may not be supervised by human beings. However, attackers, knowledgeable of how AIs are taught, can use that process to warp the behavior of an AI.

The point is this. We will use AIs even more than we do now, and that fact will mean that there will be more ways to attack us and our systems.

All of this sounds grim, doesn't it? In some ways, it is grim. The tools of progress often cut both for and against us. It is just as true for the situation we have now, and it remains true for the future.

Chapter Notes

1. "New ransomware employs TOR to stay hidden from security," Alex Hern, The Guardian, accessed 12 December 2022, http://www.theguardian.com/technology/2014/jul/25/new-ransomware-employs-tor-onion-malware
2. "';–have I been pwned?" Havelbeenpwned.com, accessed 13 December 2022, https://haveibeenpwned.com/

Index

Printed in the United States
by Baker & Taylor Publisher Services